FOREWORD

OF ALL OUR NEIGHBOUR WORLDS, the planet Mars is probably the most fascinating. In many ways it is not so very unlike the Earth, and there has always been the suggestion that it might support life. Yet until the last few years, our ideas about it were badly wide of the mark.

By now space-ships have reached Mars, and sent back information direct from the surface. Therefore, this seems to be the moment to produce what I hope is a comprehensive guide, outlining what we thought about Mars in what may be called the 'canal period' and coming on to modern times.

My most grateful thanks are due to Dr. Garry Hunt, who has read through the whole of the manuscript and made many invaluable suggestions—though I must stress that any errors or deficiencies are my responsibility, not his!

Also, I am most grateful to Lawrence Clarke, for his expert line drawings; to John Bunn, who helped so much in seeing the book through the press; and to Paul Doherty, for allowing me to use his splendid drawings of Mars. Also, I must thank the authorities at the Jet Propulsion Laboratory, California, for permission to use the Mariner and Viking pictures of Mars, without which this book would have seemed very bare indeed.

Finally, there is Michael Foxell, of Lutterworth Press, who must surely be an author's ideal publisher, and whose help and encouragement have been unfailing.

PATRICK MOORE

Selsey. *April 1977*

GUIDE
TO
MARS

Chapter One

MARS AS A WORLD

That Mars is inhabited by beings of some sort or other we may consider as certain as it is uncertain what those beings may be.

SO WROTE THE AMERICAN ASTRONOMER Percival Lowell in his book *Mars and its Canals*, published in 1906. Let it be said at once that Lowell's opinions were not shared by all astronomers of the time, and there were many who were decidedly sceptical about the brilliant-brained Martian canal-builders who had covered their planet with a network of artificial waterways. Yet the idea of Mars as an inhabited world was by no means discounted, and even today, when rocket vehicles have landed there, it remains probably the most fascinating of all the worlds in the Solar System.

Events have moved very quickly in recent years, and so far as Mars is concerned we have had to do some drastic re-thinking. In 1964 I remember writing an article in which I made a dozen positive statements, every one of which was fully supported by the best available evidence and every one of which has since been proved to have been wrong. Mars is not the sort of world we believed it to be, and each new mission raises a host of unexpected puzzles.

The Vikings of 1976 have been no exception. For the first time, it has been possible to receive intelligible messages from the Martian surface; some of them are clear-cut, but others are frankly difficult to interpret. The canals have vanished into the realm of science fiction, but the problem of 'life or no life?' is still with us. What the Vikings have told us has been spectacular enough, but the last word remains to be said.

Quite apart from the possibility of life, admittedly in lowly form, Mars is of special interest to us because it is bound to be the first world to be reached beyond the Earth-Moon system. Before the Space Age, Venus was regarded as a serious rival, but we have now learned—rather to our disgust—that Venus is intolerable by any standards, so that Mars is the only planet to hold out serious hope. It is no longer heresy to talk about Martian colonies, even though anything of the kind lies well in the future, and it is not likely that 'the first man on Mars' has yet been born.

What I propose to do, in the present book, is to give an account of Mars as we believe it to be today. New findings are coming to hand all the time, but it does seem that we have learned the basic essentials, even though we cannot yet interpret them as well as we would like to do. By this I mean that when I make positive statements in 1977, I am not likely to be as wrong as I was in 1964. First, however, let me give a few facts and figures, so that Mars can be put in its proper place in the Solar System.

Everyone knows that the Earth is a planet, moving round the Sun at a mean distance of 93,000,000 miles in a period of one year. Almost everyone knows that the Sun is an ordinary star, but it is not so generally appreciated that 93,000,000 miles is not very far on the scale of the universe. The nearest star beyond the Sun lies at a distance of over 24 million million miles, so that even light, moving at the rate of 186,000 miles per second, takes over four years to do the journey. In astronomical parlance, the nearest star (Proxima Centauri, a faint Red Dwarf too far south to be seen from England) is more than four light-years away. Most of the other stars are much more remote. Even Barnard's Star, another Red Dwarf which is distinguished by the fact that it seems to be attended by a couple of planets, is six light-years away. The distance of Sirius is 8½ light-years; Vega in Lyra 27 light-years, Rigel in Orion about 900 light-years, and so on. Look at Rigel tonight, and you will see it not as it is now, but as it used to be in the time of William the Conqueror. And when we consider other star-systems, we find that the distances have to be reckoned in millions, hundreds of millions, and even thousands of millions of light-years. Once we look beyond our own local area, our view of the universe is bound to be very out of date.

Things are different within the Solar System, which is ruled by the Sun, and whose main members are the nine planets (Fig. 1)—Mercury and Venus, closer to the Sun than we are; Mars, Jupiter, Saturn, Uranus, Neptune and Pluto further away. We see the Sun after a lapse of a mere 8½ minutes, and

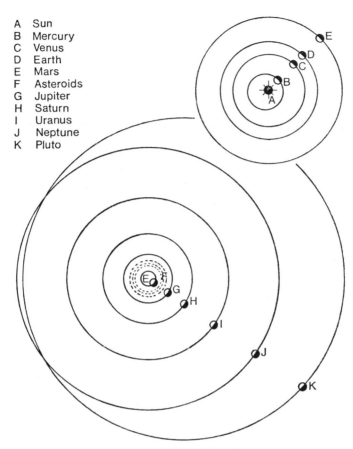

A   Sun
B   Mercury
C   Venus
D   Earth
E   Mars
F   Asteroids
G   Jupiter
H   Saturn
I   Uranus
J   Neptune
K   Pluto

Fig. 1. Plan of the Solar System

light (or radio waves) can reach us from Neptune or Pluto in about five hours. So far as rockets are concerned, our useful investigations are limited strictly to the Sun's kingdom. True, at least one of our probes—Pioneer 10, which by-passed Jupiter in late 1972—is on its way out of the Solar System for good, but there is not the slightest chance of keeping in touch with it. If we are ever to contact beings in other planetary systems, we must do so by some method which is so far removed from our current knowledge that we cannot even speculate about it—and we may be as far away from interstellar contact as King Canute was from television. The only loophole is to pick up radio transmissions from some remote planetary system, but this is a very long shot indeed.

I suppose that the burning question is always: "What are the chances of finding intelligent life within range of us?" So far as the Solar System is concerned, I am afraid that the answer must be 'Nil', and to drive this home it may be as well to give a lightning survey of the other planets, after which it should be a distinct relief to come back to Mars.

Reckoning outward from the Sun, we come first to Mercury, which has a diameter of around 3000 miles, and takes only 88 days to complete one circuit. It is never brilliant in our skies, because with the naked eye it is visible only on relatively rare occasions either low in the west after sunset or low in the east before dawn. One rocket probe, Mariner 10, has flown past it, and has sent back pictures of a crater-scarred surface which is very like that of the Moon. But Mercury has a negligible atmosphere; the temperatures are extreme, and as a potential colony it may safely be ruled out.

Venus is very different. It is about the same size as the Earth, and at its distance from the Sun of 67,000,000 miles it might be expected to be reasonably welcoming. Before the Space Age we knew very little about it, because it is permanently covered with a dense, cloudy atmosphere which our telescopes cannot pierce. There were suggestions that the surface might be largely ocean-covered, in which case primitive life might have appeared there—just as happened in the warm oceans of Earth thousands of millions of years ago.

Then, in the 1960s, came rocket probes which showed that this attractive picture was very wide of the mark. The radar mapping of the 1970s showed that there are craters on the surface, and in 1975 the Russians achieved a notable triumph by soft-landing two vehicles and obtaining one picture from each. Instead of being a friendly place, Venus has turned out to be what can only be called a psychedelic planet. The atmosphere is composed mainly of carbon dioxide, while the clouds contain quantities of sulphuric acid; the ground atmospheric pressure is 90 to 100 times that of the Earth's air at sea-level; the temperature is over 900 degrees Fahrenheit. Anyone incautious enough to go to Venus and step outside his (or her) space-craft would at once be poisoned, squashed and fried, quite apart from the corrosive effects of sulphuric acid. It is not an inviting prospect.

Number 3 in the planetary sequence is the Earth, attended by its satellite, the Moon. I do not propose to say a great deal about the Moon here, but I cannot gloss over it completely, because of inevitable comparisons with Mars; after all, both have cratered surfaces, together with high mountains and deep valleys. But the two worlds differ in many respects, and comparisons should not be taken too far.

The Moon is our companion in space; its mean distance from us is rather less than a quarter of a million miles, and its diameter (2160 miles) is more than one quarter of that of the Earth. I suspect that the Earth-Moon system should be regarded as a double planet rather than as a planet and a satellite, but the Moon has only 1/81 of the Earth's mass, and this means that its gravitational pull is too weak for it to retain an atmosphere. To call the Moon 'an airless world' is substantially correct, and there is no longer any doubt that it has been sterile throughout its long history. Rocks and surface materials brought home by the Apollo astronauts and by the Russian automatic probes have shown absolutely no trace of any organic material, and the strict quarantining of returned samples (including astronauts!) was abandoned as unnecessary after the first couple of expeditions. The same cannot be said of Mars; but of this, more anon.

There has been tremendous controversy about the origin of the Moon's walled formations, some of which are well over 100 miles in diameter. Broadly speaking, what we have to decide is whether they are of external origin, or whether they are due to internal forces within the Moon itself. According to the first theory, the craters were produced by plunging meteorites, and at any rate there can be no doubt that impact craters exist there, if only because there are some on the Earth. The Arizona Crater, not far from the Lowell Observatory at Flagstaff, was undoubtedly produced by a meteoritic fall in prehistoric times. Alternatively, it may be that most of the Moon's craters are 'volcanic', using the term in a very wide sense. Everyone has his own ideas, and mine are clear-cut; I believe that vulcanism was the main crater-building process, with impact playing a minor rôle. I have discussed this in detail elsewhere,* but I must refer to it here, because it is highly relevant when we come to consider the craters on Mars.

There is another important point, too. Originally it was thought that the Earth and the Moon used to be one body, and that the Moon was literally thrown off because of the rapid axial rotation. This idea was proposed by G. H. Darwin (son of the great naturalist) and was popular for many years before the mathematicians attacked it and more or less destroyed it. It is now generally believed that the two worlds have always been separate, but recently there have been suggestions of an original union between the Earth, the Moon and Mars—so that when Mars was thrown off, the Moon was left in between as a sort of casual droplet. I doubt whether this theory has many supporters, though it cannot be absolutely ruled out.

According to current thinking, the planets were built up by accretion from a 'solar nebula' or cloud of material; the process was a gradual one, but at least we know, with fair certainty, that the age of the Earth is of the order of 4,700 million years (or 4·7 æons, one æon being equal to a thousand million years). Analysis of the lunar samples shows that the age of the Moon is approximately the same, and no doubt this is also true of Mars. The fact that Earth, Mars and the Moon have evolved differently is because their masses are different. In particular, the Earth has retained a dense atmosphere, Mars a thin one, and the Moon virtually none at all.

Mars is the outermost of the inner or terrestrial group of planets, all of which are of modest size and have solid globes. Beyond we come to the minor planets or asteroids, all of which are below 1000 miles in diameter, and which may be either the remnants of a larger disrupted planet or (more probably) débris left over when the main planets were formed. There is a distinct chance that Phobos and Deimos, the two dwarf attendants of Mars, are captured asteroids rather than bona-fide satellites, though again this is something which lacks positive proof.

Next come the four giant planets (Fig. 2): Jupiter, Saturn, Uranus and Neptune, which are completely different in nature. Jupiter, much the most massive of them, moves round the Sun at a mean distance of 483,000,000 miles, and is so large that

* *Guide to the Moon* (Lutterworth Press, 1976).

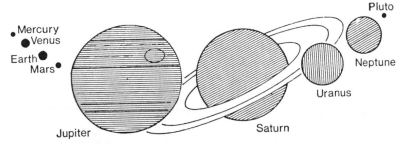

Fig. 2. Comparative sizes of the planets

its huge globe could swallow up more than a thousand bodies the volume of the Earth. It has a gaseous surface, and is now thought to have a solid core which is overlaid by layers of liquid hydrogen. In any case, it and the other giants are so unlike Earth or Mars that they need not concern us further for the moment. Yet we must not forget their numerous satellites, some of which are large. Jupiter has four really sizeable satellites, Saturn one and Neptune one. Titan, No. 6 in the Saturnian family, is almost certainly larger than Mercury, and may not be a great deal smaller than Mars; according to one estimate its diameter is 3500 miles, though admittedly this may be rather too high. Moreover, Titan has an atmosphere whose ground pressure is approximately ten times that at the surface of Mars, though the atmosphere is of different type and there is no prospect of our finding any advanced life-forms there. Titan, remember, is bitterly cold, quite apart from its other disadvantages.

Finally we come to Pluto, which is generally regarded as the outermost planet, even though for some years around 1989, the time of its next perihelion passage—that is to say, its closest point to the Sun—it will actually be nearer to us than Neptune. Pluto is an enigma. Logically it ought not to be there; it seems to be smaller than Mars, and it has a curiously eccentric orbit which takes it round the Sun only once in 248 years. From it, the Sun would look only the same size as Jupiter does to us, though the light-intensity in daytime would still be quite strong.

Probes have been sent out to the further reaches of the Solar System. Pioneer 10 by-passed Jupiter in December 1973,

17

sending back dramatic pictures as well as a tremendous amount of miscellaneous information, and has now started a never-ending journey into the space between the stars; it will never come back, and we have seen the last of it. Pioneer 11 made its pass of Jupiter a year later, and is now on its way to Saturn, which it will fly by in 1979. By then two Voyagers will be en route first to Jupiter (1979) and then Saturn (1981), possibly going on to Uranus (1985). It is all most intriguing, but because of the immense distances involved these journeys take a very long time, and any ideas of sending manned expeditions are highly premature. Whether such flights can ever be achieved remains to be seen.

In short: Venus is lethal in every way, the giants are not solid in an Earth-like sense, and the satellites (apart from Titan) are virtually devoid of atmospheres. So too is Mercury. In our search for what we may term 'local life', only Mars remains.

To show how ideas have altered in our own era, it is amusing (and instructive) to look back at the Guzman Prize, which was announced on 17 December 1900. This was quite substantial— a hundred thousand francs—and was to be given to the first man who was able to establish contact with a being living on another world. But in the terms of the citation, Mars was specifically excluded, because it was thought that contacting the Martians would be much too easy!

The Guzman Prize remains unclaimed, and there are, alas, no Martians. Luckily, the Red Planet's interest remains, and there can be no world more intriguing for man to explore.

Chapter Two

MARS IN THE SOLAR SYSTEM

MEN USED TO BELIEVE that the planets must move in perfectly circular paths or orbits. This was because the circle was regarded as the perfect form—and surely nothing except utter perfection could be allowed in the heavens? This may not seem a very scientific way of looking at things, but in bygone ages it carried a great deal of weight.

I do not want to delve too deeply into history, but since Mars comes very much into the story it is worthwhile digressing for a few moments. Ptolemy of Alexandria, the last great astronomer of antiquity, followed a scheme according to which the Earth lay at rest in the centre of the universe, with everything else moving around it: the Moon, Sun, planets and, at the very edge, the sphere of the fixed stars. Ptolemy, who was an excellent mathematician, knew that the planets do not move across the sky at unchanging speeds, and accordingly he adopted a rather cumbersome system; a planet moved in a small circle or epicycle, the centre of which itself moved round the Earth in a perfect circle. This sounds artificial, but it did fit the facts as far as Ptolemy knew them. The scheme is always known as the Ptolemaic System, though Ptolemy did not actually invent it.

Ptolemy died in or about A.D. 180, and for centuries after his death very few scientists dared to question his model of the universe. The Church, in particular, was strongly against any heretical proposal to dethrone the Earth from its proud central position. The real battle started in the fifteenth century, with the publication of a book by a Polish canon who is always remembered as Copernicus, though his proper name was Mikołaj Kopernik. Copernicus disliked the Ptolemaic System because it was over-complicated, and his remedy was to remove the Earth from the central site and put the Sun there instead. This would mean relegating the Earth to the status of an ordinary planet, no more important, basically, than Mars.

Church reaction was predictably violent, and Copernicus

only escaped severe condemnation because he was wise enough to postpone publication of his book until he was dying. Subsequently the controversy became overheated by any standards, and one rebel, Giordano Bruno, was burned alive in Rome, in 1600, partly because he refused to believe that the Sun goes round the Earth.

Meantime, a curious character had come into the story. Tycho Brahe was a Danish nobleman, who had a remarkable and picturesque life. This is no place to describe it in detail, or to recount the numerous true stories about him—though I cannot resist mentioning that during his student days he had part of his nose sliced off in a duel, and made himself a new one out of gold, silver and wax! He was a superb observer, and between 1576 and 1596 he worked away on his island observatory in the Baltic, preparing a catalogue of star-positions which was far better than anything previously drawn up. I am sure I need not stress that so far as naked-eye observers are concerned, the stars keep to the same relative positions for long periods (many lifetimes), because they are so far away that their individual or 'proper' motions seem very slight. Go back to the days of the Crusades, the Roman Occupation or even the Trojan War, and you would find that the constellations would look virtually the same as they do today. Orion, the Great Bear and the rest would be sensibly unaltered. Neither will they have changed perceptibly by, say, A.D. 2200. It is only our near neighbours—the members of the Solar System—which seem to wander about.

Tycho, unfortunately, had no telescopes. The first definite astronomical use of the telescope dates only from 1609, which was eight years after Tycho died. So the great Danish observer had to depend upon measuring instruments used with the eye alone, and under the circumstances he was amazingly successful. When he left Denmark in a huff, after quarrelling with practically everybody, he took with him a mass of observations. In addition to his star catalogue, he had made very exact studies of the movements of the planets, particularly Mars.

Tycho settled in Bohemia as Imperial Mathematician to the Holy Roman Emperor, Rudolph II, who was a gloomy, astrology-ridden individual, and whose régime was one of unmitigated disaster. It was while in Bohemia that Tycho called in, as his assistant, a young German named Johannes Kepler; and when Tycho died suddenly, in 1601, Kepler came into possession of all the priceless observations.

This was a great opportunity. Kepler, unlike Tycho, believed in the theory of a central Sun, and with Tycho's work he believed that he could prove his case. It took him a long time, and it was Mars which provided the clue. If Mars moved round the Sun, then so must the Earth—but no circular orbit seemed to fit the observations.

Kepler persisted, and blind alley followed blind alley. At last he stumbled upon the truth. Mars does indeed go round the Sun, but its path is not circular; it moves in an ellipse. Once this had been realized, everything clicked neatly into place. The main battle was over, though the last shots remained to be fired; Galileo, the first great telescopic observer, was tried and condemned by the Roman Inquisition in 1633 because of his refusal to accept the Ptolemaic System, and it was not until 1687, with the publication of Isaac Newton's *Principia*, that all serious opposition ceased. But it was Kepler who had provided the essential evidence.

I say that Kepler 'stumbled on the truth' because he had had the solution in his grasp for some time without realizing it; a great deal of moral courage was needed to break free from the age-old concept of circular orbits. Incidentally, it was pure luck that Tycho's main observations had been concentrated on Mars, because the path of Mars is much more eccentric than those of, say, Venus or Jupiter. Venus really does have an orbit which is almost circular, and if Mars had behaved similarly the solution would have been much harder to find. On the other hand, one must not belittle Kepler. He had the utmost faith in Tycho's observations, and if he had been less confident about them he would inevitably have failed.

In fact, Mars has an orbital eccentricity of 0·093, as against 0·017 for the Earth and only 0·004 for Venus. This means that its distance from the Sun ranges between 128,410,000 miles at perihelion or closest point out to 154,860,000 miles at aphelion or furthest recession—a difference of over 25,000,000 miles, as against less than four million miles for the Earth. The mean distance between Mars and the Sun is therefore rather less than 142,000,000 miles. It is, incidentally, only the second nearest planet to the Earth. It may come within 35,000,000 miles of us, but this is more than ten million miles further than Venus at its closest.

When well placed, Mars is the brightest object in the entire sky apart from the Sun, the Moon and Venus. It stands out not only because of its brilliance, but because of the strong red colour which led the ancients to name it in honour of the God of War. At its faintest it may drop down to near-equality with the Pole Star, and at such times it is not easy for the beginner to identify it, because it is frankly inconspicuous (Fig. 3).

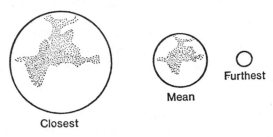

Fig. 3. The varying apparent size of Mars

Mars takes 687 Earth-days to go once round the Sun. Obviously it is at its best when opposite to the Sun in the sky, but these 'oppositions' do not occur every year; the mean interval between them is 780 days, though this is not absolutely constant. To show what happens, let us refer to the diagram, in which the orbits of the two planets are drawn to the same scale.

It will be convenient to start on 10 August 1971, because on that date Mars was at opposition (M1 in Fig. 4a). It was also practically at perihelion, so that it was almost as close to the Earth as it can ever be; it was glaringly conspicuous, and was visible throughout the hours of darkness. One year later the Earth had completed a full journey round the Sun, and had arrived back at its starting-point (E1), but Mars, moving more slowly in a larger orbit, had not had time to do so. Its mean orbital velocity is only 15 miles per second, as against 18½ miles per second for the Earth. By August 1972 Mars had only reached position M2, and was almost behind the Sun. The Earth had to catch it up, and it achieved this only on 25 October 1973, when there was another opposition (Fig. 4b). Throughout

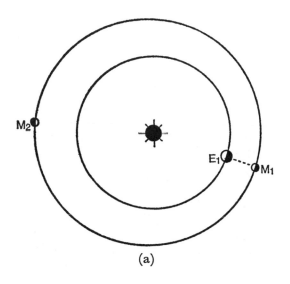

(a)

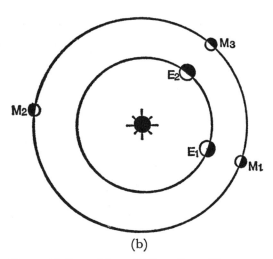

(b)

Fig. 4. (a) Oppositions of Mars. Opposition occurs
with the Earth at E1 and Mars at M1. A year later
the Earth has returned to E1, but Mars has only
travelled as far as M2. (b) The Earth has therefore
to catch Mars up, and the next opposition occurs
when the Earth has reached E2 and Mars has
reached M3

1972 Mars was badly placed for observation, and small telescopes could show practically nothing on its disk.

This is why oppositions occur in cycles of just over two years. The next was that of 15 December 1975; 1976 was missed, and so was 1977, but Mars will again reach opposition on 22 January 1978, after which we must wait until 25 February 1980.

Note, too, that not all oppositions are equally favourable. That of 1980, for instance, will take place with Mars near

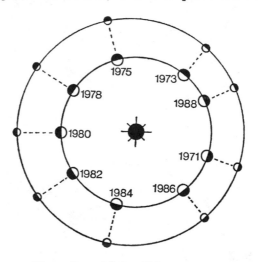

Fig. 5. Oppositions of Mars, 1971–88

aphelion, and the minimum distance from us will never be reduced to as little as 60,000,000 miles (Fig. 5). To make matters worse from the point of view of northern-hemisphere observers, favourable oppositions always take place when Mars is south of the equator of the sky, so that it is low down as seen from Europe or the United States. To see Mars at its very best, you must go south—to Australia, for instance, or South Africa.

Like all the planets, Mars shines only by reflecting the light of the Sun. At opposition, the whole of the sunlit hemisphere faces us, and Mars is full, but well away from opposition the disk no longer appears circular. Mars shows a distinct phase, and at times only 86 per cent. of the sunlit hemisphere is turned in our direction. Sketches made at the telescope have to take this into account, because if the disk is drawn as circular, even when Mars shows a phase of below 90 per cent., errors are bound to be introduced. Only when the planet is reasonably close to opposition can the careful observer afford to draw the disk as perfectly circular. The disks given on page 190 may be of help.

This may be the moment to introduce the Martian seasons, which are all-important in any consideration of the surface conditions. As a preliminary, I must say something about our own seasons, which are due in the main not to our changing distance from the Sun, but to the tilt of the Earth's axis, which is $23\frac{1}{2}$ degrees to the perpendicular. In Figure 6a, the Earth is shown in two positions. During northern summer, the northern hemisphere is tilted toward the Sun; the north pole is in continual sunlight, while the south pole has no daytime at all.

Northern Summer   Northern Winter

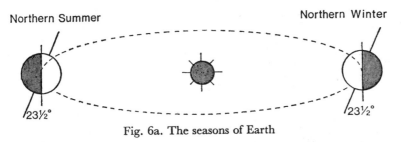

Fig. 6a. The seasons of Earth

Six months later, the situation is reversed. The Earth is at perihelion during northern winter—only $91\frac{1}{2}$ million miles from the Sun, as against $94\frac{1}{2}$ million in late June—but the difference is not enough to have any marked effect, and in any case the greater amount of ocean in the southern part of the Earth has a stabilizing influence.

Not so with Mars, which has no oceans at all, and where the range between perihelion and aphelion is so much greater (Fig. 6b). At the present epoch, the axial tilt is virtually the same as ours (24 degrees), and, as with us, southern summer occurs near perihelion. This means that on Mars the southern summers are shorter but hotter than those in the northern hemisphere, while the southern winters are longer and colder.

It is one of Kepler's fundamental laws that a planet moves fastest when at its closest to the Sun; this is a fundamental traffic rule of the Solar System.

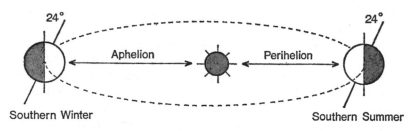

Fig. 6b. The seasons of Mars

The effects of this situation are far-reaching, and I will refer to them over and over again during the course of this book. In short: the southern hemisphere of Mars has climates which are more extreme than those of the north, and the differences are quite marked. It is also worth noting that during close oppositions it is the south pole which is tilted in our direction, so that before the Space Age the southern hemisphere was the better-mapped of the two.

Though Mars has such a long 'year', its 'day' is much the same as ours—to be precise, a little more than half an hour longer: 24 hours 37 minutes 22·6 seconds, a period which has been measured very exactly from Earth. Since the Viking landings, a Martian day has become known as a 'sol', and this is the term which I propose to use from now on. Needless to say, a sol indicates a complete rotation—approximately $12\frac{1}{4}$ hours of daylight followed by $12\frac{1}{4}$ hours of darkness at the equator at the time of the equinoxes. The solar day (or, rather, solar sol) is 24 hours 39 minutes 35 seconds. This is the mean interval between successive transits of the Sun across the meridian as seen by an observer on Mars.

The near-equality between Earth day and Martian sol means that we can draw up a calendar which is not entirely unfamiliar. The lengths of the seasons, for instance, can be worked out easily enough:

Martian Season	*Length*	
	Days	Sols
Southern spring (northern autumn)	146	142
Southern summer (northern winter)	160	156
Southern autumn (northern spring)	199	194
Southern winter (northern summer)	182	177
	687	669

Timekeeping is not going to be a real problem for future colonists (as it would be on Venus, where a rotation period or 'day' is actually longer than the revolution period or 'year', leading to a very peculiar state of affairs). A Martian clock will no doubt be divided into 24 hours, though each will be slightly longer than an Earth hour. So far as a calendar is concerned—well, there have already been plenty of suggestions. My own idea is to divide up a Martian year into 18 months, each of 37 sols. This makes 666 sols. We need 669, so an extra sol can be tacked on to Months 6, 12 and 18, giving them 38 sols instead of 37. This should work reasonably well, though there will have to be some kind of adjustment to allow for the fact that the revolution period of Mars is not exactly 687 days, but 686 days 23 hours 52 minutes 31 seconds according to Earth reckoning. The problem of naming our eighteen Martian months is not, I feel, something which need be considered yet awhile, though no doubt it will produce the usual tedious international squabbles when it eventually has to be tackled.

Incidentally, the Martian pole star is not the same as ours. The Earth's axis points northward in the sky to a position very close to Alpha Ursæ Minoris—Polaris, which seems to remain almost stationary with everything else moving around it. (I am, of course, talking from the viewpoint of a northern-hemisphere observer. Australians and South Africans can never see Polaris, and they have no bright south polar star, the

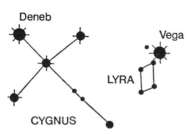

Fig. 7. Position of Deneb, the nearest bright star to the Martian North Celestial Pole

nearest naked-eye object being the obscure Sigma Octantis.) Mars is well served. The north polar star there is Deneb or Alpha Cygni (Fig. 7), which is of the first magnitude and will make a splendid marker for future explorers. This is just as well, because the magnetic field of Mars is so weak that conventional compasses will not work.

However, Deneb will not retain its title indefinitely, because the axial tilt of Mars is not constant. The same is true of the Earth, because of the phenomenon known as the precession of the equinoxes. Our world is not a perfect sphere; it is somewhat flattened at the poles, and the equatorial diameter is about 26 miles greater than the polar (7926 miles, as against 7900). Mars is flattened to a slightly greater extent, since the discrepancy amounts to more than forty miles, and the shape is rather less regular. Effects of this lead to a precession which makes the axial inclination range between 35 degrees and only 14 degrees. Obviously the shift is slow, but it may have very important effects upon the Martian climate, as we will see later.

The virtual absence of a magnetic field brings us on to the question of what Mars is like inside its globe. The mean density is less than that of the Earth; the so-called 'specific gravity' is only 3·94, so that Mars 'weighs' 3·94 times as much as an equal volume of water would do. For the Earth, the figure is 5·5, indicating that there is a much more substantial core made up of heavy substances such as iron. The Moon, with a specific gravity rather less than that of Mars (3·34) is thought to have a core with a diameter of about 360 miles, over which lies a region of 'partial melting' which is in turn overlaid by the mantle and finally by the relatively thin crust, which goes down to only about 30 to 40 miles below the lunar surface.

Mars is likely to be rather more lunar than terrestrial in structure, and the space-probe results of the past few years have at least given some valuable clues. As befits a world with only one-tenth the mass of the Earth, Mars may have built up rather quickly from the material of the solar nebula, and one estimate suggests that it was formed in only about 100,000 years, which is a mere tick on the cosmical clock. The outer layers then solidified to form rocks of various types (Fig. 8). The lighter rocks floated to the top to make a crust, while the lower ones

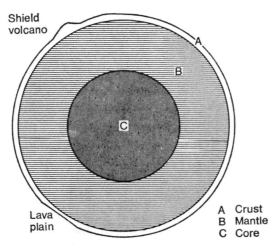

Fig. 8. Probable internal structure of Mars

became hot, and the heavy substances such as iron sank to make up a core. The internal heating may have been due largely to radioactivity. Some elements, such as uranium, decay spontaneously into lighter elements; uranium, for instance, ends its career as a form of lead, and this phenomenon causes a surprising amount of heat over a long period when considerable quantities of radioactive elements are involved. At any rate, it does seem almost certain that Mars has a substantial iron-rich core, though it is smaller than the Earth's both relatively and absolutely. This may account for the apparent absence of a measurable magnetic field—even though the origin of planetary magnetic fields is still rather obscure, and it would be dangerous to be too positive.

Next, there is the all-important question of surface gravity. Again we may expect a value intermediate between those of the Earth and the Moon. The surface gravity of a planet (or a satellite) depends upon a combination of its mass and its diameter; the greater the mass and the smaller the diameter, the greater the surface pull. The reason why diameter is important is because a body behaves as though all its mass were concentrated at a point in its centre—and the further away you are from the centre, the weaker the gravitational tug. A very good case of this is given by the giant planet Uranus, whose mass is 14½ times that of the Earth, and whose diameter is almost 30,000 miles. Anyone standing on the surface of Uranus would therefore be much further away from the centre of the planet than anyone standing on the surface of the Earth, and the surface pull would be unexpectedly slight. Despite the mass-difference, the surface gravity on Uranus is practically the same as that on the Earth. (Of course, I am taking an impossible case, because Uranus has a gaseous surface and nobody could possibly stand upon it. However, I think that the principle is clear.)

With the Moon, the surface gravity is approximately one-sixth of that on Earth, so that an astronaut seems to have only one-sixth of his 'normal' weight, and when he walks about everything appears to happen in slow motion. Most people will have seen this on television during the Apollo missions. Fortunately, it has been found that the sensation is neither uncomfortable nor dangerous, and the situation will be quite satisfactory on Mars, where the surface gravity is approximately one-third of our own. (Note, by the way, that the surface gravity on Mercury is almost exactly the same as that on Mars. Mercury is much the smaller of the two planets, but it is also much denser, with a larger heavy core.)

Escape velocity also depends upon mass, but here the diameter is not so important when we are considering bodies of planetary size. By now 'escape velocity' has become part of our everyday language, but it may be as well to give a brief account of it, because it is so vital in any discussion of conditions on Mars.

Throw an object upward, and it will rise to a certain height, stop, and fall back. Throw it harder, and it will travel further upward before returning. If it were possible to throw the object upward at a velocity of 7 miles per second, or about 25,000 m.p.h., it would not fall back at all; the Earth's gravity would be unable to retain it, and the object would escape into space. Therefore, 7 miles per second is the Earth's escape velocity. Actually, any body moving through the dense lower atmosphere at 7 miles per second would be destroyed by frictional heating against the air-particles, which is one reason why the space-guns so beloved of past science-fiction writers cannot work, but again the principle should be clear.

The Moon has an escape velocity of $1\frac{1}{2}$ miles per second, and for Mars the value is 3·1 miles per second. (For Uranus, it happens to be 14 miles per second.) And this has an effect upon the density of atmosphere remaining.

Air is made up of vast numbers of tiny particles, all moving around at high speeds. Obviously, a particle moving outward from the Earth with a velocity of 7 miles per second will escape. Fortunately, the Earth's air is made up of particles which cannot attain this velocity, and there is no danger of our being left gasping like a goldfish which has been removed from its tank. But the Moon, with its weak gravitational pull, has been unable to hold on to any atmosphere it may once have had, which is why it is so unfriendly and sterile. Mars, predictably, is an intermediate case. It would be expected to have a thin atmosphere—and this is true, though admittedly the atmospheric density is much less than we believed before the first rocket vehicles sent us back information from close range.

According to modern ideas, there was a period in the Earth's early history when there was no air at all. The original atmosphere, made up chiefly of hydrogen, escaped; hydrogen is the lightest of all the elements, and its atoms and molecules are the most difficult to retain. Later, a secondary atmosphere was formed from gases set free from the interior. No doubt the same sequence of events occurred on Mars, and the Vikings have provided additional evidence that this really did happen.

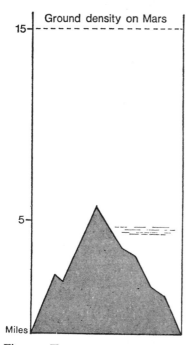

Fig. 9. Tenuous atmosphere of Mars. The pressure of the Earth's atmosphere at a height of about 15 miles above sea level is approximately equal to the pressure at the surface of Mars

We now know that at the present moment, the Martian atmosphere is so thin that the ground pressure is everywhere below 10 millibars, as against almost 1000 millibars at sea-level on the Earth. This means that the Martian atmosphere is no denser than the Earth's air at a height of at least 15 miles above the ground (Fig. 9), and we would be unable to breathe it even if it were made up of pure oxygen (which it is not). From this, we can show at once that there can be no liquid surface water, and there are no seas, lakes or even ponds—though things were certainly different in the past, and will probably be different again in the future.

The early observers had other ideas. With their telescopes, they drew the markings on Mars, and believed that they were looking at oceans, continents, islands and straits. It was only with the development of more modern-type instruments that the truth began to emerge. But I am running ahead of my story, so let us now turn back the pages to these far-off times even before the telescopic exploration of Mars began.

Chapter Three

EYE AND TELESCOPE

NOT EVEN THE MOST myopic observer can overlook Mars when it is best placed. Quite apart from its red colour, it is so brilliant that in bygone days it has even caused panics—as in 1719, when it was widely mistaken for a crimson comet which was hurtling toward us and might destroy the Earth. Obviously, then, it must have been known since the dawn of human history, and it was certainly one of the first bodies to be identified as a 'wanderer' rather than a true star.

In ancient times astrology, the superstition of the stars, was taken very seriously indeed, and Mars was regarded as having an evil influence. This is no place to discuss astrology, which is, after all, strictly for the credulous; to modify a famous remark said to have been made by the Duke of Wellington, "anyone who believes in astrology will believe in anything". But it is understandable that our remote ancestors should have formed a deep distrust of Mars, because it is so red. Red indicates blood, and blood means war. What could be more natural than to name the planet after the War-God? The Greek war-god was Ares, and the study of Mars is still known officially as 'areography', though by now the term seems pedantic, and I doubt whether it will survive. 'Martian geography' is so much more explicit, though technically deplorable.

The Greeks were well aware of the unstarlike nature of Mars, and they studied its movements closely. Yet they were not the first to do so. Apparently the Egyptians knew the planet as Harmakhis or Har decher—that is to say, the Red One, while the Chaldæans called it after Nergal, the Babylonian god of war. The first precise observation of its position seems to date from 272 B.C., just over half a century after the death of Alexander the Great; on 17 January of that year it is recorded that Mars passed very close to the star Beta Scorpii. Later on, the Arabs and Persians knew Mars as Mirikh, indicating a torch, while in India it was Angaraka, from *angara*, a burning coal.

In pre-telescopic times nothing could be learned about Mars itself. The only kind of observation possible was that of position-measuring, and by the time that telescopes arrived on the scene the way in which Mars behaved was very well known—thanks principally, but by no means entirely, to Tycho Brahe. Those who are sufficiently interested can easily plot the changing rate and direction of Mars as it moves across the night sky. Near opposition it will seem to move backwards, or east to west, because the Earth is 'catching Mars up' and passing it. Figure 10 should make this clear; the backwards motion is termed 'retrograding', and was one of the phenomena which

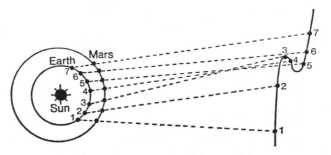

Fig. 10. Retrograding of Mars. It is obvious that as the Earth 'overtakes' Mars, the planet will seem to reverse its direction for an appreciable period before resuming its normal direction of motions

showed from the outset that Mars could not move round the Earth in a circular path at unchanging speed.

The telescope was invented in the early years of the seventeenth century. There can be no doubt that a Dutch spectacle-maker named Hans Lippershey had developed the instrument by 1608, and he is widely assumed to have been the first, though scientific historians have their doubts, and some of them suspect that the principle dates back half a century further. (Since spectacles had been in use for so many years, I have always been surprised that the discovery of the telescope was so long delayed.) Thomas Harriott, one-time tutor to Sir Walter Raleigh, made a telescopic map of the Moon in 1609, and he was only one of several pioneers; but in those early days pride of place must go to the great Italian, Galileo Galilei, who made a telescope for himself and began astronomical observing during the winter of 1609–10. At once he made a series of spectacular discoveries. The planet Jupiter was found to be attended by four moons or satellites of its own; the phases of Venus were obvious; the Moon was a world of mountains and craters; there was something very odd about the shape of Saturn, and the Milky Way was made up of countless stars. Galileo was quick to publish his discoveries, some of which caused a great deal of controversy. Remember, the argument between the supporters of the Ptolemaic and the Copernican systems was still raging, and the Church was deeply involved. Everything that Galileo saw through his telescope strengthened his belief in the theory that the Earth and the other planets move round the Sun. The phases of Venus were of special significance, because on the old Ptolemaic pattern Venus would always be seen as a crescent—never as a half or full disk.

There were two planets which puzzled Galileo. One, of course, was Saturn, which seemed at first to be a triple object, though later Galileo found that the two side-members had disappeared. As we now know, this aspect was due to the famous system of rings, which could be seen in Galileo's low-powered, imperfect telescope, but not clearly enough for him to decide what they were. (Their temporary disappearance was due to the fact that in 1612 the rings were tilted edgewise-on to the Earth, as happens regularly; the last occasion was in 1966, the next will be in 1980.) Mars was certainly single, but it did not seem to be quite circular, and on 30 December 1610 Galileo wrote to his friend P. Castelli: "I ought not to claim that I can see the phases of Mars; however, unless I am deceiving myself, I believe I have already seen that it is not perfectly round." He was correct, and the fact that he detected the phases shows how good an observer he was.

What he could not do, of course, was to see any surface features on Mars, and this brings me to a point which must be stressed at the outset. Mars is not an easy object to study. It is a very difficult one, and small telescopes will show little on its disk. The modestly-equipped amateur observer is at a marked disadvantage, and unless a powerful telescope is available Mars will defy all attempts to tackle it.

The reason is straightforward enough. Mars is a small world, and it is by no means on our cosmical doorstep, since it can never approach us much within 35,000,000 miles. This is around 150 times as far away as the Moon, and a comparison may be helpful. Even when at its nearest, Mars has an apparent diameter of only 25·7 seconds of arc, and it would fit comfortably into one of the Moon's craters, say Plato. No wonder that high magnifications are needed to show it well.

However, the early telescope-users were not to be deterred, and they did their best. So far as is known, the very first telescopic sketch of Mars was made in 1636 by an Italian amateur, Francisco Fontana, who lived in Naples. The drawing (Fig. 11)

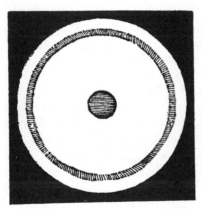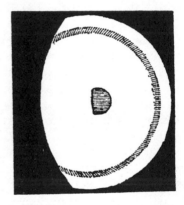

Fig. 11. Drawings of Mars by F. Fontana: (*left*) 1636, (*right*) 1638

is hardly of great scientific value, but Fontana's own words are worth noting: "The form of Mars was observed to be perfectly spherical. In its centre was a dark cone in the form of a very black pill. The disk was of many colours, but appeared to be flaming in the concave part. Except for the Sun, Mars is much the hottest of all the stars." A second drawing, made on 24 August 1638, shows the same 'pill', with a grossly exaggerated phase.

Fontana cannot be blamed for misinterpreting what he saw. The 'pill', of course, was purely an optical effect due to the poor quality of his telescope (he recorded a similar spot on Venus), and the 'many colours' were equally spurious. At least he had made an effort to be constructive, which was all to the good.

Passing over one or two earnest but frankly unsuccessful efforts to record detail on Mars during the years following Fontana's sketch, we come to Christiaan Huygens, who is an important figure astronomically. His telescopes were strange by modern standards, since they were small-aperture refractors of immensely long focal length, but he was certainly the best observer of his time, and he is credited with several outstanding triumphs. For instance, he discovered the true nature of Saturn's ring system, and he also found Titan, the largest and brightest member of the Saturnian family of moons. (Huygens' pioneer work in other fields, notably clockmaking, need not concern us here.) Huygens made a drawing of Mars in 1656, but it was on 28 November 1659 that he recorded the first true feature: the triangular dark marking which has been named, at various times, the Atlantic Canal, the Kaiser Sea, the Hourglass Sea, the Syrtis Major, and now, since the space-probe flights, Syrtis Major Planitia.

The drawing was made at seven o'clock in the evening. The Syrtis is unmistakable; true, its size is exaggerated, but of its identity there can be no doubt at all. It is still there, and is the most obvious dark feature on the whole of the disk. Huygens' sketch (Fig. 12) gives extra proof, if proof were needed, that the Martian markings are permanent rather than being mere clouds, as with Venus. Moreover, one can check back and see whether the Syrtis Major really was in the position in which Huygens drew it on the disk. The agreement is excellent.

Huygens looked at the Syrtis Major over a period of hours (at least, so we must assume), and he saw that it was being carried slowly across the disk. The direction would be from right to left as shown in the sketch, because the orientation is with the south at the top.* Obviously, the rate at which a

* Most astronomical telescopes give an upside-down image, and until recently all planetary and lunar drawings and photographs were given with south at the top and north at the bottom. To the regret of old-fashioned observers such as myself, the trend nowadays is to have north at the top. This was the subject of a heated discussion some years ago at a meeting of the International Astronomical Union, the controlling body of world astronomy, and in a final vote the 'south-toppers' such as myself were heavily defeated!

Fig. 12. Drawing of Mars by Huygens, 28 November
1659. The Syrtis Major is unmistakable

marking shifts is a clue to the planet's rotation period. Huygens
made no mistake, and in his diary for 1 December 1659 he
recorded that "the rotation of Mars, like that of the Earth,
seems to be in a period of twenty-four hours". This is a mere
half-hour or so wrong, and Huygens could scarcely have done
any better.

During his career Huygens made various other drawings of
Mars, again showing the Syrtis Major, but he did not have the
monopoly. Another great pioneer was Giovanni Cassini,
Italian by birth but who spent much of his life in France, and
became the first Director of the Paris Observatory. In 1666
(actually before he went to Paris) Cassini made several
observations of Mars, and although his drawings were not so
good as Huygens' he found that the markings came back to the
same positions on the disk forty minutes later each day—so that
after 36 to 37 days they returned to the same places at the same
times. It followed that the rotation period of Mars must be
slightly longer than that of the Earth, and Cassini arrived at a

value of 24 hours 40 minutes, which is practically correct. There was some argument about this, and a 13-hour period was initially preferred by the telescope-maker Campani and his colleagues in Rome, but before long Cassini's results were confirmed.

The next great step was the discovery of the white polar caps, which looked so much like snowy or icy mantles that their nature was not seriously questioned—particularly as they were found to wax and wane according to the Martian seasons. One man who saw them was Giacomo Maraldi, Cassini's nephew, who recorded them in 1704 and again at the favourable opposition in 1719 (Fig. 13). Maraldi gave the rotation period

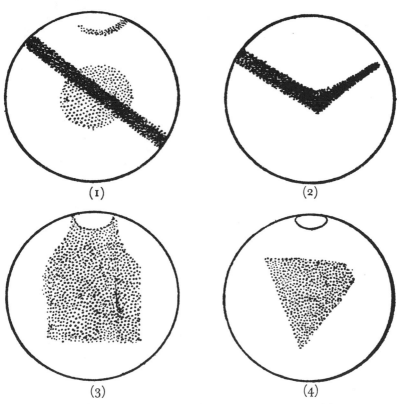

Fig. 13. Drawings of Mars by Maraldi, 1719. (1) 13 July. (2) 19 August, 25 September and 28 October. (3) August–October. (4) 5 August and 16 October

as 24 hours 39 minutes, which was correct to within two minutes.

We now come to Sir William Herschel, who is one of the giants of astronomical history. He was Hanoverian by birth, but spent most of his life in England. In 1781, during one of his 'reviews of the heavens', he discovered the planet Uranus; he catalogued large numbers of star-clusters and nebulæ; he found that some of the double stars in the sky are physically-associated or binary systems rather than mere line of sight effects, and he was the first man to draw up a reasonably good model of the shape of the star-system or Galaxy. King George III gave him a modest pension, and he received every honour that the scientific world could bestow, including a knighthood and the Presidency of the newly-formed Royal Astronomical Society (then the Astronomical Society of London). Herschel was also the foremost telescope-maker of his time, and his most powerful instrument, a reflector with a mirror 49 inches in diameter, was in a class of its own; you can still see the mirror if you go to Flamsteed House, the old Royal Observatory in Greenwich Park.

Herschel's main work was in connection with the stars, but he did pay attention to Mars in the period between 1777 and 1783, using relatively small telescopes of his own manufacture. He was fully convinced that the polar caps represented thick masses of snow and ice, and he confirmed earlier suggestions that they were not centred exactly on the geographical poles. He measured the diameter of Mars, and he fixed the rotation period at 24 hours 39 minutes 50 seconds. Later, the two German observers, Beer and Mädler, re-worked Herschel's observations and gave a revised period of 24 hours 37 minutes 23·7 seconds, which was only about one second too long.

To Herschel, Mars was a world with lands and seas—and, of course, atmosphere. Yet Herschel did not believe the atmosphere to be nearly so dense as had been assumed by Cassini and others, and he had what he regarded as observational proof. In its journey among the constellations, Mars sometimes passes in front of a star, and hides or occults it; even if there is no actual occultation, a 'near miss' would mean that the star would pass behind any widespread shell of Martian atmosphere, and would be noticeably dimmed. On 26 and 27 October 1783 Herschel watched two faint stars as they passed within a few seconds of arc of the Martian limb, and found that they were not affected at all. Presumably, then, the atmosphere of the planet was not really extensive. Again Herschel was right. Yet another of his researches involved the flattening of the globe of Mars, which is more marked than that of the Earth, and he fixed the axial inclination as being at an angle of 28 degrees. As we have noted, the true value is 24 degrees.

Herschel's observations of Mars virtually ended in 1783, but two years later a long series was begun by Johann Hieronymus Schröter, chief magistrate of the German town of Lilienthal, near Bremen. Schröter was in regular correspondence with Herschel, and one of his telescopes (probably the best one) was of Herschel's manufacture, but the two men were working along very different lines. Schröter was concerned entirely with the Solar System. He was the first really great observer of the Moon, and he also studied the planets as well as taking an active part in more speculative programmes. He was president of an organization formed in 1800 specifically to search for small planets moving round the Sun in orbits between those of Mars and Jupiter, and one of them, Juno, was found by his assistant, Karl Harding, working at the Lilienthal observatory.

I have always had a great admiration for Schröter. He has been accused (rightly) of being a clumsy draughtsman, and also (wrongly) of being inaccurate. It has even been claimed that his telescopes were of poor quality. We can never find out, because his observatory was destroyed in 1813 during the Franco-German war, and the brass-tubed telescopes were plundered by the French soldiers under the impression that they were gold, but it would be most unwise to suggest that the

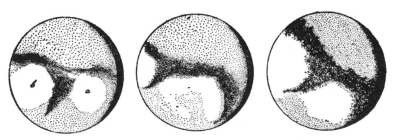

Fig. 14. Drawings of Mars by Schröter, 8 December 1800

Herschel reflector was anything but excellent. In any case, Schröter made some good drawings (Fig. 14), and if he had had the inclination he could easily have combined them into a proper Martian map. He made various measurements (diameter, axial inclination, polar flattening and so on) which were better than Herschel's and his work represented a distinct observational advance. Theoretically he was not so close to the mark, and, surprisingly, he wrote that he had "never observed with certainty any completely fixed dark patches which, like our seas and lakes, would have a lower reflecting power". He believed that the dark patches were nothing more than clouds in the Martian atmosphere.

It is hard to see how he can have fallen into such a trap, particularly as he made many observations which showed the same features time and time again. He also detected all the phenomena of the polar caps which are so well known today—the variations in brightness and extent, the somewhat irregular outlines, and the seasonal cycle, but he misinterpreted them hopelessly.

Before coming on to Beer and Mädler, who produced the first comprehensive chart of Mars, I must pause to say something about Honoré Flaugergues, a French amateur who is best remembered for being the discoverer of the brilliant comet of 1811. Flaugergues had his private observatory at Viviers, and observed Mars during several oppositions, including that of 1813. He was convinced that the markings were variable, and, unlike Schröter, believed them to be true surface features. He also realized that during Martian spring and early summer, a polar cap will shrink very quickly. Flaugergues assumed that the cap must be a thick layer of ice and snow, and he went on to draw a very remarkable conclusion:

"If the melting of the polar ices on Mars is much more prompt and much more complete than with our own terrestrial ice-caps, most of which persist up to the heat of summer, it seems therefore that the heat on Mars is greater than on Earth, though because of the planet's greater distance from the Sun it ought to be less in the ratio of 43 to 100. This is an extra reason to add to those which have made the most skilful physicists believe that the rays of the Sun do not in themselves cause heat, but are only the indirect cause of heat."

In other words, Mars has a hotter climate than that of the Earth! One can follow Flaugergues' reasoning, and he was not alone in his views. The idea of Martian inhabitants was taking root. Indeed, it was not new. Even Herschel had stated that in his opinion all the planets were inhabited, and he even believed that there were intelligent beings living in a cool, pleasant region beneath the hot surface of the Sun. Mars, with its marked resemblance to the Earth, was an obvious candidate.

Then, from 1830 to 1841, came the observations by Wilhelm Beer and Johann von Mädler, who worked together at Beer's private observatory outside Berlin. They made an interesting partnership. Beer—brother of Meyerbeer, the composer—was a banker by profession; Mädler was the main observer, and the telescope used was a modest 3¾-inch refractor. Despite this

1837

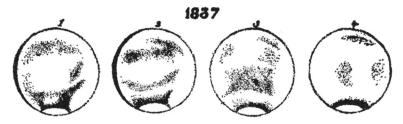

Fig. 15. Drawings of Mars by Beer and Mädler, 1837

small aperture, Beer and Mädler compiled a map of the Moon which remained the best for half a century, and still bears comparison with modern outline charts. They also drew up a map of Mars, and selected a dark feature to represent longitude zero. Relatively speaking, their map of Mars was not so good as their lunar chart, but it was certainly better than anything which had been produced before, and under the circumstances it was a remarkable achievement. Quite apart from this, they revised and improved all the earlier physical measurements— including the rotation period, which, as we have seen, they worked out to an accuracy of less than one second.

Mädler left Berlin in 1840–41 to become Director of the new Dorpat Observatory in Estonia (then, as now, under Russian rule). Neither he nor Beer did much more lunar or planetary work, which was a pity. It must be admitted that the following two decades were rather barren, apart from some good drawings made in the mid-1850s by a British amateur, Warren de la Rue, who was also a pioneer photographer. De la Rue used a 13-inch equatorial reflector, which must have been optically very good.

The main feature on many of de la Rue's drawings, such as that of 20 April 1856, is the Syrtis Major, which extends across much of the disk and has a pronounced 'tail'. The southern cap is also well shown, and there are various other recognizable details. He also showed whiteness in the far north. It had already been pointed out (in 1853, by the famous French astronomer François Arago) that the southern cap should show greater variations of area than the northern, because the climates there are more extreme.

In 1858 some detailed drawings of Mars were made by Angelo Secchi, a Jesuit who became famous for his pioneer work in the field of stellar spectroscopy. He recorded the Syrtis Major, and it was he who called it the 'Atlantic Canal', though the name could not have been worse chosen—and there is no association with the celebrated canals which became so controversial in after years.

Up to this time it had been tacitly assumed that the dark patches on Mars were seas, while the ochre tracts represented continents. One man who did not agree was Emmanuel Liais, who was trained in Paris but was then invited by the Emperor of Brazil to become Director of the Rio de Janeiro Observatory, and who spent most of the rest of his life in South America. Liais made some rather desultory observations of Mars, but his important contribution was the suggestion that the dark regions were likely to be vegetation-tracts rather than oceans. Liais' theory was published in 1860. Secchi did not agree, and two years later wrote that "the existence of seas and continents . . . has today been conclusively proved". Even at that stage, Mars was providing plenty of scope for argument.

Better telescopes and better techniques led steadily to more accurate maps of Mars. Sir Norman Lockyer, another pioneer of astronomical spectroscopy, produced some good drawings during the favourable opposition of 1862, and agreed with Secchi that the 'green' areas were oceanic. Frederik Kaiser in Holland and Camille Flammarion in France were other skilful observers; Kaiser gave the rotation period as 24 hours 37

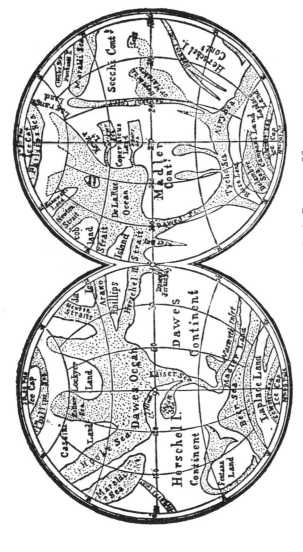

Fig. 16. Map of Mars, by R. A. Proctor, 1867

minutes 22·62 seconds. Meanwhile, spectroscopy had been coming to the fore, and at first it seemed to support the idea that the atmosphere of Mars was decidedly damp. Flammarion, for one, was convinced that the dark regions were due to water in some form, though he did wonder whether it could be in a kind of intermediate state, neither pure liquid nor pure vapour.

Next, there was the question of naming the Martian features. This was something which had never been seriously tackled, and the names used by different observers were unofficial. The lead was taken by Richard A. Proctor, a British amateur who was a noted 'popular' writer but also a clever astronomer in his own right. In 1867 Proctor produced a map in which he gave the Martian features names in honour of famous observers —Cassini Land, Fontana Land, Mädler Continent, Arago Strait and so on (Fig. 16). His system was followed by other British observers, such as N. E. Green and the Rev. W. R. Dawes.

Promptly and predictably, the storm broke. Proctor's map was criticized as being inaccurate, and there was some truth in this; he was also attacked for having selected names which favoured British astronomers. (Accusations of the same type were levelled at the Russians nearly a century later, when they obtained the first pictures of the far side of the Moon and produced names such as the Moscow Sea and the Soviet Mountains.) Various modifications were introduced, but the whole system was finally thrown overboard in favour of a new one by Schiaparelli. I may be running ahead of my story, and if so I apologize; but it is interesting to compare the rival nomenclatures, and so let us look at two maps, one based on Proctor's system and the other on Schiaparelli's. The identifications are as follows:

Proctor (and Green)	*Schiaparelli*
Beer Continent	Aeria and Arabia
Herschel II. Strait	Sinus Sabæus
Arago Strait	Margaritifer Sinus
Burton Bay	mouth of the Indus
Mädler Continent	Chryse
Christie Bay	Auroræ Sinus
Terby Sea	Solis Lacus
Kepler Land and Copernicus Land	Thaumasia
Jacob Land	Noachis and Argyre I

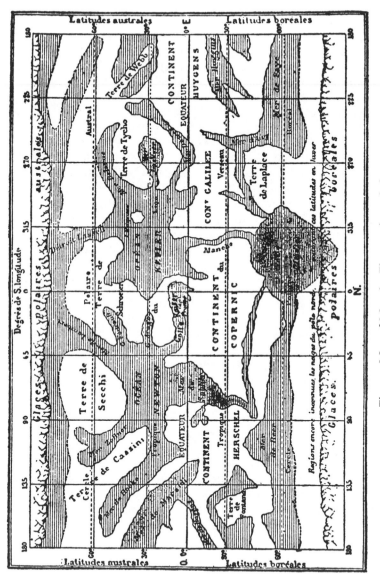

Fig. 17. Map of Mars by Flammarion, 1876

Proctor (and Green)	*Schiaparelli*
Phillips Island	Deucalionis Regio
Hall Island	Protei Regio
Schiaparelli Sea	Mare Sirenum, Lacus Phœnicis
Maraldi Sea	Mare Cimmerium
Hooke Sea and Flammarion Sea	Mare Tyrrhenum and Syrtis Minor
Cassini Land and Dreyer Island	Ausonia and Iapygia
Lockyer Land	Hellas
Kaiser Sea (or the Hourglass Sea)	Syrtis Major

Which system do you prefer? I admit to having a feeling of nostalgia for Proctor's; after all, it is the kind of nomenclature which has been accepted for the Moon. On the other hand one is bound to admit that Schiaparelli's, based partly on geography and partly on mythology, is the more scientific, particularly now that we know so much more about the nature of Mars; Arago Strait is not a strait, Phillips Island is not an island, and the Hourglass Sea is not a sea—it is a lofty plateau sloping off to either side. Of course, Schiaparelli's system has had to be revised too. Lacus Phœnicis—the Phœnix Lake—has proved to be a towering volcano. Note that Chryse, the 'Golden Land' where Viking 1 made its epic touchdown in July 1976, was the old Mädler Continent. One wonders what Johann Mädler would have thought.

Such was the situation at the end of what we may term the middle period in Martian history. The movements, the shape and the main surface features of Mars were well known; the caps, varying seasonally, were assumed to be icy or snowy; the atmosphere was thought to be reasonably dense, though less so than that of the Earth; the dark regions were either oceans (as most astronomers believed) or else vegetation tracts, presumably filling old sea-beds. The temperature was expected to be little, if at all, lower on average than that of Earth, and the idea of inhabitants was certainly not ruled out.

The stage was set for one of the most memorable years in the story of Mars: 1877.

Chapter Four

THE STORY OF THE CANALS

To the early observers, Mars was an attractive world. It was also a mysterious one, because so little could be seen upon it apart from dark patches and white polar caps. Speculation was rife, and it is worth looking back to some comments by Bernard de Fontenelle, secretary of the French Academy of Sciences, in 1688. Fontenelle regretted that Mars had no satellites—at least so far as he knew; the two dwarf attendants were not discovered until almost two centuries later—and then went on as follows:

"We have seen phosphorescent materials, either liquid or dry, which upon receiving light from the Sun absorb it, so that they can shine brilliantly when in shadow. Perhaps Mars has great, high rocks, naturally phosphorescent, which during the day can store up light, emitting it again during the night. Nobody can imagine a pleasanter scene than that of rocks illuminating the whole landscape after sunset, and providing a magnificent light without inconvenient heat. In America we know that there are many birds which are so luminous that in darkness we can read by their light. How do we know that Mars does not have a great number of these birds which, when night comes, scatter on all sides and make a new day?"

This is at least far more inviting than the bloodthirsty Martians conjured up later by H. G. Wells and his imitators. In fact, the idea of hostile inhabitants did not occur to anyone until modern or near-modern times, and although men such as William Herschel believed firmly in the habitability of Mars they did not go so far as to speculate what the inhabitants might be like. Obviously the first requirement was to communicate with them, and various interesting suggestions were made. Johann von Littrow, who became Director of the Vienna Observatory in 1819, proposed lighting vast fires arranged in geometrical patterns to attract the attention of the Martians, who would presumably understand the message and make a suitable reply. Another bright idea was to dig wide trenches in

49

the Sahara Desert, and provide a sort of mathematical code. The climax was reached in the mid-1870s by Charles Cros, an enthusiastic Frenchman, who put forward the scheme of building a large burning-glass which could focus the Sun's light and heat on to a Martian desert, scorching the sand there; by swinging the glass around it would be possible to write words on the surface of Mars. I have often wondered what words he intended to write, but the plan never progressed even as far as the drawing-board stage. Monsieur Cros was apparently most upset at the general refusal to take him seriously.

If we could not signal to the Martians, could we hope to see any evidence of their handiwork? This brings us straight on to the famous observations made in 1877 by Schiaparelli, setting off a violent argument which was not finally ended until the flights of the Mariners. It was Schiaparelli who made the first detailed studies of the features which he called *canali*, or channels, but which have been immortalized as the Martian canals.

Giovanni Virginio Schiaparelli was born in Piedmont in 1835, and graduated from Turin University. In 1862 he was appointed Director of the Brera Observatory in Milan, which was equipped with a fine $8\frac{3}{4}$-inch refractor. His interests were many (for instance, he carried out pioneer work in connection with comets and meteor streams), but for the moment we must confine ourselves to his work on Mars, which began with the 1877 opposition. The actual opposition date was 5 September, and Mars was practically at perihelion, so that it was to all intents and purposes as close to us as it can ever be. Schiaparelli was a skilled observer, and the Milan skies were clear (much clearer than they are today). He therefore decided to compile a new map of Mars.

The chart which he produced was certainly much better than any of its predecessors, and in the main it stands up quite well to the modern results. He also revised the nomenclature; out went Beer Continent, Lockyer Land and Dreyer Island, while in came Aeria, Hellas and Iapygia. For a while the two systems ran in parallel, but eventually Schiaparelli's prevailed, and I propose to use it from now on, albeit with a slight feeling of regret.

The most striking feature, however, was the detection of

very fine, regular lines running across the reddish-ochre deserts. They were, Schiaparelli believed, unlike anything else in the Solar System, and he was frankly taken aback. In a later article he summarized his ideas about them, so let us keep to his own words:

"All the vast extent of the continents is furrowed upon every side by a network of numerous lines or fine stripes of a more or less pronounced dark colour, whose aspect is very variable. They traverse the planet for long distances in regular lines, that do not at all resemble the winding courses of our streams. Some of the shorter ones do not reach three hundred miles; others extend for many thousands, occupying a quarter or even a third of the circumference of the planet. Some of these are very easy to see, especially the one designated by the name of Nilosyrtis. Others in turn are extremely difficult, and resemble the finest thread of a spider's web drawn across the disk. They are subject to great variations in breadth, which may reach 120 to 180 miles for the Nilosyrtis, while others are scarcely 20 miles broad ... Their length and arrangement are constant, or vary only between very narrow limits ... The canals may intersect among themselves at all possible angles, but by preference they converge toward the small spots to which we have given the name of lakes. For example, seven are seen to converge in Phœnicis Lacus, eight in Trivium Charontis, six in Lunæ Lacus and six in Ismenius Lacus."

Perhaps the most remarkable thing about the canal network as described by Schiaparelli was that it seemed to follow a definite pattern. There was nothing haphazard about it. Either the canals followed great-circle tracks across the planet, such as the Phison, or else they were gently curved, such as the Nilosyrtis. Whether curved or not, they ran from dark area to dark area; there was not a single case of a canal breaking off abruptly in the middle of an ochre tract. Altogether, Schiaparelli recorded forty canals during the 1877 opposition.

It has often been said that Schiaparelli was not the first to see the canals, and that earlier observers had drawn some of them. This is basically correct. What seems to be a 'canal' is shown on one of the drawings made by Beer and Mädler, and there are streaks on the sketches made by observers such as Lassell, Lockyer, de la Rue, Secchi, Kaiser and others. The

Rev. W. R. Dawes, in 1864, produced drawings on which there are streaks which would certainly have been called canals if the term had been invented then so far as Mars was concerned. But the aspect as shown by Schiaparelli was entirely different, and it opened up a new train of thought.

The next opposition was that of 1879. Mars was rather further from perihelion, but conditions were still good, and Schiaparelli made the most of them. He recovered the old canals, and added new ones. There was something more: single canals could be abruptly replaced by double ones, a phenomenon which became known as 'gemination' or twinning. To quote Schiaparelli again:

When a gemination occurs, "the two lines follow very nearly the original canal, and end in the place where it ended. One of these is often positioned as exactly upon the former line, but it also happens that the two lines may occupy opposite sides of the former canal, and be located upon entirely new ground. The distance between the two lines differs in different geminations, and varies from 370 miles and more down to the smallest limit at which the two lines may appear separated in large visual telescopes—less than an interval of 30 miles." According to Schiaparelli's observations, a canal which appeared single one night might well be double the next (Fig. 18).

When Schiaparelli first published his results there was a good deal of scepticism, which was understandable. Nobody else saw the canals in 1877, and although C. E. Burton in Ireland, using a 6-inch refractor and an 8-inch reflector, made a few sketches in 1879 which showed significant streaks, full confirmation of the canal network was not forthcoming for some time. Succeeding oppositions passed by with Schiaparelli still obtaining results which differed from those of other observers. For that matter, no two observers seemed to show Mars in the same guise, and a selection of maps of the time underlines what I mean. Compare charts such as the Knobel, using an 8-inch reflector in 1884; Lohse, also in 1884; and Schiaparelli at the previous opposition. It is hard to credit that they represent the same planet.

Schiaparelli was not in the least deterred. Also, it had to be borne in mind that the oppositions of the early 1880s were less favourable than those of 1877 and 1879, because Mars was

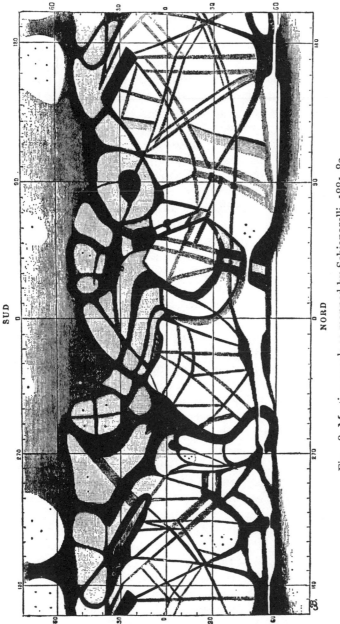

Fig. 18. Martian canals as mapped by Schiaparelli, 1881–82

furthur from perihelion and its apparent diameter was smaller. Therefore the lack of confirmation was hardly unexpected, and Schiaparelli was convinced that the 'canali' were true channels, carrying away the flood-water produced by the melting of the ice-caps at the poles. At this stage he seems to have regarded them as natural, though afterwards he became more inclined to the view that they were artificial. He even wrote that "Their singular aspect has led some to see in them the work of intelligent beings. I am very careful not to combat this suggestion, which contains nothing impossible."

Finally, in 1886, came what appeared at that time to be full verification of the canal network. Using the powerful 30-inch refractor at the Nice Observatory in France, two professional astronomers—Perrotin and Thollon—published a chart which was as strange-looking as anything which Schiaparelli had produced. Subsequently, canals became thoroughly fashionable. It would be tedious to list all the observers who recorded them, though I must mention A. Stanley Williams in England and François Terby in Belgium.

Schiaparelli himself ended his main series of observations in 1890, because of eyesight trouble; he made some drawings afterwards, but was reluctant to publish them because he did not trust them. (Sad to say, he lost his sight completely a year or two before his death in 1910.) Meanwhile, two Americans had come very much to the fore: William H. Pickering and Percival Lowell. Pickering, a professional astronomer attached to Harvard, was an expert observer—he discovered the ninth satellite of Saturn, Phœbe—and was a lunar and planetary specialist. It is true that his ideas tended to be rather unconventional, and he tended to believe that some of the allegedly variable dark spots on the Moon were due to swarms of insects, but of his skill at the eye-end of a telescope there could be no doubt at all. He made numerous drawings of the Martian canals from 1892 onward, and it was he who gave the name of 'oases' to the dark patches in which the streaks intersected. He found that some of the oases were centres of radiating canals, and saw at least six issuing from the prominent dark area known as the Trivium Charontis, then regarded as a deep depression. More importantly, he discovered that there were canals crossing the dark regions as well as the bright deserts, which to all intents and purposes gave the death-blow to the idea that the dark areas were watery.

Yet the central figure in the Mars story became, and remained, Percival Lowell, who began his career as a diplomat and then decided to devote his life to astronomy. In 1894 he founded an observatory at Flagstaff, in Arizona, principally to study Mars; he equipped it with a fine 24-inch refractor, and he worked away untiringly until shortly before his sudden death in 1916.

Flagstaff was not chosen lightly. Lowell knew that as well as having a large telescope he had to have the clearest possible skies, and the field was wide open. Schiaparelli had used a relatively small instrument, but Lowell was well off, and he was able to go more or less where he liked. Tests showed that Flagstaff was exceptionally well-favoured climatically, and so to Flagstaff he went. The Lowell Observatory quickly became famous, and it still is. Nowadays there are many powerful telescopes there, and the 24-inch refractor has been relegated to a minor rôle, but it is undoubtedly one of the best instruments of its type. I have used it myself on many occasions, so I can speak with experience.

Lowell is remembered as the man who perfected the canal network, and who claimed that Mars must undoubtedly be inhabited by beings capable of building a planet-wide irrigation system, drawing water from the icy caps and transferring it to the populated regions closer to the equator. He was wrong on both counts. Yet we must also remember that he achieved a great deal of immensely valuable work in other fields, and it was he who made the calculations which led to the tracking-down of the ninth planet, Pluto; the discovery was made at Flagstaff, though admittedly not until years after Lowell's death. There can be no doubt that Lowell was a great scientist, who will always retain an honoured place in the history of astronomy.

During his lifetime he was joined, at various periods, by many skilled observers, including Pickering. Canals aplenty were recorded: geminations, variations, oases—in fact, phenomena of unique kind. Lowell was in no doubt about the reality of the canals. In his book *Mars and its Canals* (1906) he wrote that "the Martian canals when well seen are not at the limit

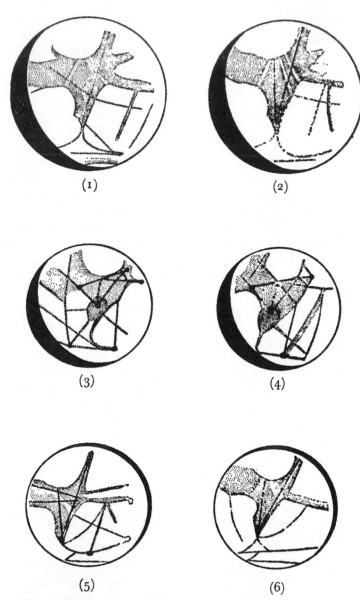

Fig. 19. Drawings of the Syrtis Major region by Lowell. (1) 2 October 1896, 14.25. (2) 2 October 1896, 15.07. (3) 9 October 1896, 17.22. (4) 9 October 1896, 18.40. (5) 11 January 1897, 05.01. (6) 12 January 1897, 04.51

of visibility, but well within that boundary of doubt . . . under good atmospheric conditions the canals are comparable for conspicuousness to many of the well-recognized Fraunhofer lines, and are just as certainly there." (The Fraunhofer lines are the dark lines seen in the spectrum of the Sun; many thousands have been mapped with absolute precision.)

Lowell's geometrical canal network could not, he maintained, be natural. Therefore the planet must support an advanced civilization, though he was careful to add that "to talk of Martian beings is not to mean Martian men". To him, the inhabitants had long since renounced warfare, which he described (accurately enough!) as "a survival among us from savage times, and now affecting chiefly the boyish and unthinking element of the nation". Mars was short of water because it was more advanced in its evolution than the Earth, and had aged more quickly; therefore, the Martians were doing their utmost to extract every scrap of moisture from the only reserves left—the ice and snow in the polar caps.

Two other phenomena seemed highly relevant. When a polar cap became smaller in spring and early summer, Lowell noted a dark band round its edge, which shrank as the cap shrank; this was attributed to moisture which produced temporary lakes or marshes, which would presumably be less reflective. Secondly, he claimed that the canals and the main dark areas were affected by the shrinking of the cap. According to his theory, the released moisture swept down toward the equator, and the vegetation revived from its winter hibernation, so that a 'wave of darkening' was seen over a period of weeks or months. I will have more to say about this 'wave' later. Lowell even calculated that its rate was approximately two miles per hour. With regard to the dark areas, he wrote: "Quickened by the water let loose on the melting of the polar cap, they rise rapidly to prominence, to stay so for some months, and then slowly proceed to die out again. Each in turn is thus affected . . . One after another each zone in order is reached and traversed, till even the equator is crossed, and the advance invades the territory of the other side. Following in its steps, afar, comes its slower wane. But already, from the other cap, has started an impulse of like character that sweeps reversely back again, travelling northward as the first went south. Twice each Martian year is the main body of the planet traversed by these waves of vegetal awakening, grandly oblivious to everything but their own advance. Two seasons of growth it therefore has, one coming from its Arctic, one from its Antarctic, zone, its equator standing curiously beholden semestrally to its poles."

Lowell did not insist that a canal must be a channel of open water. Evaporation would be an obvious hazard, for instance. It seemed more likely that the narrow central canal was flanked to either side by strips of cultivated land, and the oases presumably represented population centres. But so far as Lowell was concerned, everything was dependent upon one claim: Mars was inhabited, and the canal network was artificial.

Not surprisingly, speculation followed speculation. Dykes, huge walls, and even strings of giant pipes were suggested. In a book called *The Riddle of Mars*, published in 1914, a writer named C. E. Housden went into great detail about the nature and arrangement of the pumping stations needed to transfer water from the poles to the equator, via the canals. Housden proved to his entire satisfaction that the water-bearing pipes must be about six feet in diameter, and were used to force the water up to high-level service reservoirs by way of the canals, from whence the water was distributed to the lands by static pressure. Coming on to more modern times, Donald Lee Cyr, in 1944, attributed the canals to fertility tracts, made by bands of creatures tracking across the surface from one oasis to another—though Cyr believed the oases to be natural craters on Mars produced by the impacts of meteorites. Cyr seems therefore to have been the first to publish a suggestion that there might be craters on Mars. He was followed by E. J. Öpik some time later, but Öpik's ideas were very different from Cyr's.

Not everyone believed that the canals were artificial. Pickering considered that they were natural cracks in the surface, through which gases could escape from the interior and which were filled with vegetation. And even in Lowell's lifetime there was strong criticism of his theories, which were regarded as extreme. One violent attack was launched by Alfred Russel Wallace, one of the pioneers of the theory of evolution, who wrote a whole book on the subject in 1907, concluding that water must be absent from the planet and that the temperature was hopelessly low. He ended with the sentence: "Mars, therefore, is not only uninhabited by intelligent beings such as Mr. Lowell postulates, but is absolutely UNINHABITABLE!" (The capitals are Wallace's.)

If Lowell's maps had been accurate, there could have been little doubt that the features were non-natural. But the crux of the whole problem was—Did the canals exist at all? Were they straight and regular; were they broader, diffuse streaks totally unlike artificial waterways; or were they nothing more than tricks of the eye?

Significantly, the maps of the network drawn by different observers did not agree at all well. Apart from the most obvious of the streaks, the 'canals' shown by Schiaparelli, Lowell, Perrotin and Thollon, Leo Brenner and others were in different positions and of different intensities. Moreover, in some charts—notably Brenner's—the complexity of the system was unbelievable by any standards. This is not to say that there was no agreement at all; there were a few canals, notably the Nilosyrtis and the Nepenthes-Thoth, which were shown by almost everyone who could see the network at all. Yet even here, some observers showed them as broad, diffuse stripes, while others preferred the spider's-web appearance drawn by Lowell and his colleagues.

N. E. Green, the pioneer English observer who had been making careful studies of Mars well before the revelations of 1877, suggested that the canals were not true streaks, but merely the boundaries between areas of different colours. Vincenzio Cerulli, an Italian enthusiast who established an observatory at his home town of Teramo, proposed that the canals were not straight and regular lines, but were made up of disconnected spots and streaks; as he pointed out, the human eye does tend to join small features together when straining to glimpse objects at the very limit of visibility. Cerulli was unprejudiced, because he could himself see what might be interpreted as canals; he had no faith in Lowell's interpretations, but neither did he dismiss the canal system as a pure illusion.

Cerulli's theory was tested in 1903 by E. W. Maunder, in England. He made some drawings of Mars, without canals, and showed them at a distance to a class of boys from the

Royal Greenwich Hospital School, telling them to make copies. When the boys did so, many of them showed sharp, linear canals, and Maunder drew the obvious conclusion. Lowell was not impressed. He dismissed the idea contemptuously as the 'small boy theory'.

For my own satisfaction, I repeated the experiment in 1950. The boys, at a Kent preparatory school, were between 10 and 13 years old—rather younger than those in Maunder's class, but of a higher educational level, and more used to drawing. The pictures shown were actual drawings of Mars made through a large telescope, but with disconnected spots and streaks put in along the alleged canal sites. Out of a total of 58 boys, 42 showed vague indications of 'something' where the canals should have been, 13 showed continuous broad, hazy strips, and only three showed Lowell-type canals. Of these three, two of the boys were notoriously inartistic and the third short-sighted. I give the results here to demonstrate that Lowell did, frankly, have right on his side when he preferred to trust trained observers rather than schoolboys!

And yet there were experienced observers who completely failed to see the canals in any form whatsoever. One of these was George Ellery Hale, planner of the great reflectors at Mount Wilson and Palomar. Another was Asaph Hall, who had discovered the two Martian satellites in 1877. Most astronomers believed Lowell to be wrong in claiming that the network was glaringly obvious, and after Lowell's death in 1916 the doubters grew in number.

Probably the most skilful planetary observer of the inter-war years was Eugene M. Antoniadi, a Greek-born astronomer who went to France and spent most of his life there. For a time he worked with Camille Flammarion, the great popularizer of astronomy (and, let it be added, an excellent observer in his own right). Later Antoniadi was able to make extensive use of the 33-inch refractor at Meudon, near Paris, which again I know well and which is a superb instrument. With it, Antoniadi drew up the best map of Mars to be compiled before the space-probe era. He could see 'canals', using the term in its most general sense, but to him they were by no means artificial in appearance, and he had no patience whatsoever with Lowell's work, either observational or theoretical. Finally, in 1930, he published a

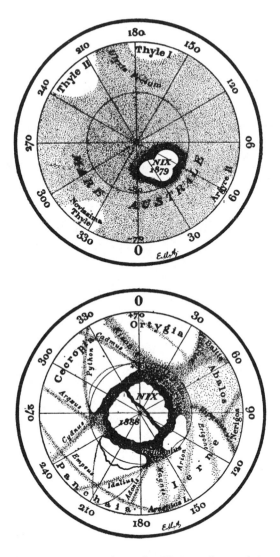

Fig. 20. Polar regions as shown by Flammarion and Antoniadi

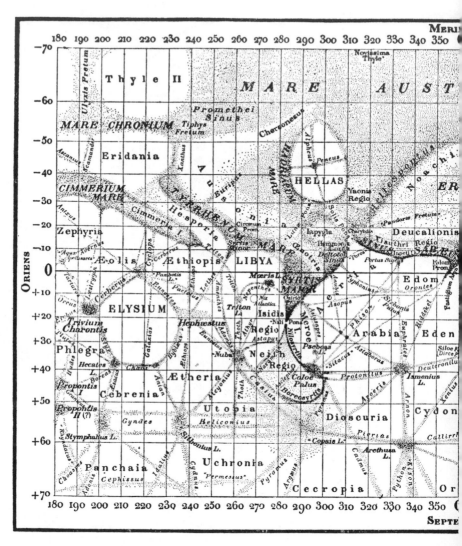

Fig. 21. General map of Mars, drawn in 1901 by Flammarion and Antoniadi

IES

10 20 30 40 50 60 70 80 90 100 110 120 130 140 150 160 170 180

-70

Argyre II

R A L E

Thyle I

-60

·Dia·

Pallouri Fretum

Charitum Prom.

MARE CHRONIUM

-50

ARGYRE I

Obydia Regio

Bosporus Gemmatus

Aonius Sinus

Phaethontis Electris

-40

Hipparus Prom.

THRÆUM MARE

Thaumasia Fœlix

Icaria

MARE SIRENUM

Atlantis II

-30

Pyrrhæ Regio

Protei Regio

Nectaris Pons

Nectar

SOLIS LACUS

Dædalia

Titanum Sinus

-20

Margaritifer Sinus

Aromatum Prom.

Aurea Chersonesus

L. Phœnicis

Memnonia

-10

Thymiamata

Oxia Palus

Hydraotes

Iamuna

Jurenis Pons

Ophir

Aromatum Prom.

Aurorae Sinus

Tithonius

Iris

Eumenides

Gorgonum

Gigas

Ammonium

Brontes

Titan

0

OCCIDENS

Snera

Tharsis

Nodus ·Gordii

Lucus Marleæ

Orcus

+10

Chryse

Gigas

Ascræus L.

Lucus Gordii

+10

Luna L.

Uranius

Nix Olympica

Amazonis

Pyriphlegethon

Brebus

+20

Indus

Hydaspes

Niliacus

Niliokeras

Nilus

Tantalus

Acheron

Lucus

Titan

+30

Deuteronius Lacus

Achillis Pons

Dardanus

Labeatis L.

Dardanus

Phlegethon

Acheron

Propontis I.

+40

Jordanis

Sirbal

MARIA ACIDALIUM

Acherial

T e m p e

Centaunus

Arcadia

Tantalus

Phlegethon

Herculis Pons

+50

a

Clarius

Castorius L.

Steuos

Propontis II (?)

+60

e

Tannis

Issedon

Mæotis Palus

Eurotas

Hebrus

Jaxartes

Cedron

Baltia

Nerigos

Illissus

Erigone

Arsenius L.

Ilias

Choaspes

+70

ygia

Abalos

Hippalus

Scandia

10 20 30 40 50 60 70 80 90 100 110 120 130 140 150 160 170 180

TRIO

E. M. Antoniadi

monumental work which contained a complete description of the Martian surface.*

Antoniadi's views were clear-cut. "Here is the actual truth of the matter. Nobody has ever seen a genuine canal on Mars, and the more or less rectilinear, single or double 'canals' of Schiaparelli do not exist as canals or as geometrical patterns; but they have a basis of reality, because on the sites of each of them the surface of the planet shows an irregular streak, more or less continuous or spotted, or else a broken, greyish border or an isolated, complex lake . . . It is necessary to add here that Schiaparelli's 'canals', which have a basis of reality, are quite different from the completely illusory 'canals' shown on the drawings of Mars . . . by Lowell and his assistants. Several authors have compared Lowell's charts of Mars sarcastically with spider's webs. Such a comparison is not inappropriate."

We now know that Antoniadi was correct in rejecting the Lowell network, but in my view he was being a little harsh, because it is now clear that Schiaparelli's canals do not exist either. Antoniadi did not hesitate to point out that Lowell had also drawn linear features upon other planets, particularly Mercury and Venus, which are equally non-existent. So far as the oases were concerned, Antoniadi rejected them because he had seen them "perfectly round even when a long way from the centre of the disk"; had they been real, they would have been foreshortened.

Antoniadi's opinions carried a great deal of weight. (He died in Occupied France during the Second World War.) But even in the years when rocket probes were being developed, the canal puzzle was still a matter of debate. Users of small telescopes often produced drawings which were very much on Lowell's pattern, and as recently as 1959 Clyde Tombaugh, who discovered the planet Pluto in 1930 while working at Flagstaff, proposed that the oases were impact craters, while the canals linking them were natural cracks filled with vegetation—not an original idea, since something of the sort had

* For some reason which I cannot explain, it was not translated into English for many years. Eventually I produced a translation, which was published—but not until 1975, by which time it was of historical interest only.

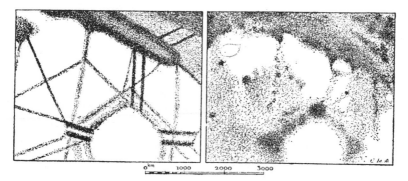

Fig. 22. Antoniadi's explanation of the canals. (*Left*) Canals in the region of Elysium, drawn by Schiaparelli between 1877 and 1890. (*Right*) The same region, drawn by Antoniadi between 1909 and 1926 with the Meudon 33-in. refractor; the hard, linear features are broken down into fine detail

been suggested long before by the Swedish physicist Svante Arrhenius, but one which was still regarded as credible.

During the 1950s, a long series of observations was carried out by Audouin Dollfus at the Pic du Midi Observatory in the French Pyrenees. The refractor there, a 24-inch, is as large as Lowell's, and the Pic is 10,000 feet high, so that the atmosphere is exceptionally steady and transparent. Broadly, Dollfus divided the linear features into three distinct classes—wide, shady bands; narrow, more regular streaks; and 'canals', thread-like, perfectly black, and artificial looking. He maintained that the latter were purely physiological, produced by the human eye, while under conditions of perfect seeing the wider bands and streaks could be broken down into separate dots and patches.

And yet . . . Consider the observation by R. S. Richardson, one of America's most respected astronomers, made on 4 June 1956. Richardson had never previously seen canals, despite all his efforts, and was frankly sceptical about them. His account is well worth quoting:

"I was on Mount Wilson working with the 60-inch reflector. My object was to take some test exposures on Mars with various filters and emulsions in preparation for observations in the fall when the planet came to opposition. Mars at the time was 75,000,000 miles away . . . As soon as Mars came into view in the eyepiece I knew the seeing must be pretty good. It was easy to hold for several seconds detail that could only be suspected before. The dark band around the south polar cap was very evident . . . A magnification of 700 made it seem as if Mars were being viewed from a space-ship. I was familiar with the appearance from my observations of 1954, but this morning the disk had a peculiar aspect I had never noticed before. The bright red regions were covered with innumerable irregular blue lines, like veins running through some mineral. Some minutes passed before it occurred to me that these markings must be canals. I was taken completely by surprise, as I had not been thinking of canals, and certainly did not expect to see them at a distance of 75,000,000 miles."

True, he added that "their appearance was not artificial, but gave the impression of being some natural feature". Observations of this sort certainly supported Antoniadi's belief that the canals had a basis of reality—at least in some cases.

It may not be out of place for me to put in a word about my own observations, which began in the 1930s, though it was not until after the war that I had the opportunity to turn really large telescopes toward Mars. I was never able to see a canal. The sketch given in Plate I was made on 28 August 1973 with the 27-inch refractor at Johannesburg, under good conditions, and there are no streaks of any kind. I make no claim to be a keen-eyed observer, but it is fair to add that many others, far more skilful than I can ever be, had the same experience. By 1973, of course, the canal question had been cleared up, but neither did I see any linear streaks earlier on, though I looked for them more times than I care to remember.

I hope you will forgive this lengthy digression into past history, because it is so deeply involved with the whole story of Martian exploration. The final solution was deferred until the Mariners of 1965 and succeeding years. Mariner 4, the first of them, sent back pictures which showed craters and rough areas, but no canals; Mariners 6 and 7 did likewise in 1969; and then, in 1971–72, came the epic flight of Mariner 9, which resulted in the first really good, high-resolution map of the Martian surface. The volcanoes, valleys and craters showed up with crystal clarity, and if there had been any trace of linear canals they could not possibly have been missed. In fact there was

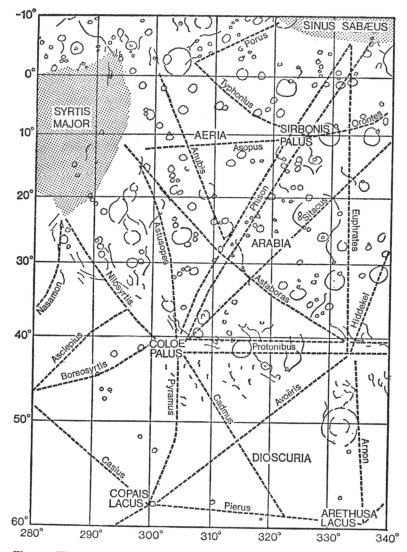

Fig. 23. The region of Aeria, Arabia and Dioscuria. The details are taken
from the maps obtained from the Mariners and Vikings, and the canal
network is superimposed. It is clear that there is no correlation whatsoever,
and that the canals do not correspond to any real features

absolutely no sign of them, and even the valleys and mountain groups did not correspond with any of the Lowell-type features.

In a very few cases there is some correlation, but in the vast majority of instances we must, I am afraid, dismiss the canals as being due to tricks of the eye. The honesty of the observers cannot be questioned; they were simply misled, and this again underlines how little we could really find out about Mars before we reached the stage of being able to send probes there.

As an epitaph, I have selected part of one of the modern maps of Mars, obtained from the Mariner 9 pictures, and super-imposed upon it the canals as they were shown by the earlier observers (Fig. 23); my main reference source was a map, published in 1910 by Flammarion and Antoniadi. I chose this region because it includes the Syrtis Major and the Sinus Sabæus, as well as some of the canals whose existence was regarded as absolutely definite—the double Phison, the Euphrates, the Arnon, the Hiddekel, the Casius and so on. It is painfully clear that none of these has even a 'basis of reality', to use Antoniadi's own term. There are vague darkish patches in the sites of the old Ismenius Lacus and Coloe Palus, but no sign of any radiating streaks, and another reported dark patch, the Siloe Fons, seems to be non-existent.

The canals have their place in history. They will never be forgotten, and most people—including myself—sincerely regret that they have been disproved. But Science is remorselessly logical, and the century-old argument is over. Nobody has ever seen a true canal on Mars; there are no canals to see.

Chapter Five

MARS BEFORE MARINER: ATMOSPHERE, CLOUDS AND DUST-STORMS

THE HISTORY OF THE exploration of Mars can be divided into several parts, each of which is reasonably well-defined. They are: (1) Very early days, pre-1830, when only the main features were known—that is to say, dark areas and white polar caps. (2) 1830 to 1877, when maps became at least reasonably accurate, and telescopes were powerful enough to show how Mars behaves. (3) 1877 to 1965, when the canal controversy raged, but modern-type techniques provided information which was thought to be reliable. (4) 1965 to the present time, when unmanned probes have been to the planet and have caused a complete change in our outlook.

In the present chapter I propose to deal with Period 3, but with only scant reference to the canals, because I have already discussed them in detail. I must also gloss over most of the eccentric ideas proposed from time to time, such as the idea that the ochre tracts might be seas and the dark areas lands. However, I cannot resist mentioning a book written by a German named Ludwig Kann, published in Heidelberg in 1901. Kann maintained that Mars is completely water-covered; the ocean is covered with floating masses of yellowish-red weed except in the dark zones, where there is no weed, and the sea-bottoms are visible through clear water. The canals are gaps in the weed caused by ocean currents . . .

In this survey it will, I feel, be appropriate to begin with what was known about the Martian atmosphere. Of its existence there had been no serious doubt since the eighteenth century. As long ago as 1784 William Herschel had written that "Mars is not without considerable atmosphere; for besides the permanent spots on its surface, I have often noticed occasional changes of partial bright belts; and also once a darkish one, in a pretty high latitude. And these alternations we can hardly ascribe to any other cause than the variable disposition of clouds and vapours floating in the atmosphere of the planet . . .

(There is) a considerable but moderate atmosphere, so that its inhabitants probably enjoy a situation in many respects similar to our own."

As we have noted, a thin atmosphere on Mars would be expected from the rather low escape velocity of 3·1 miles per second, and over the years many attempts were made to measure the ground pressure. Unfortunately all the methods used were subject to very considerable doubt, and before the Space Age there was no way round the problem. One line of research was to see how the brightness of various patches on the disk varied according to the distance from the central meridian, since a feature near the limb would be seen through a thicker layer of Martian atmosphere than one near the centre of the disk. In 1939 Gérard de Vaucouleurs, at Le Houga Observatory in France, made a long series of estimates, and came to the conclusion that the ground pressure must be about 7 centimetres of mercury, or slightly more than 90 millibars, as against an average pressure of 960 millibars at sea-level on the Earth. Various other methods were also used. Since all are now obsolete there is no point in going into details about them, but it is interesting to list some of the values derived:

1934	N. Barabaschev (U.S.S.R.)	50 millibars
1940	N. Barabaschev (U.S.S.R.)	116 ,,
1941	V. V. Sharonov (U.S.S.R.)	120 ,,
1944	N. Sytinskaya (U.S.S.R.)	112 ,,
1945	G. de Vaucouleurs (France)	93 ,,
1948	S. L. Hess (U.S.A.)	80 ,,
1951	A. Dollfus (France)	83 ,,
1960	E. Öpik (Ireland)	116 ,,

Apart from Barabaschev's early estimate, there seemed to be general consensus of opinion that the ground pressure of the Martian atmosphere was between 80 and 120 millibars—about the same as that in the Earth's air at a height of eleven miles above sea-level. Remember that Mount Everest is less than six miles high, so that even if the Martian atmosphere had been made up of pure oxygen it would still have been of little use to beings such as ourselves. Incidentally, it was obvious that a terrestrial barometer would have to be re-calibrated for use on Mars!

There was one very important conclusion to be drawn. With a ground pressure of around 85 millibars, liquid water could exist on Mars provided that its temperature did not rise much above 100 degrees Fahrenheit. Of course, few people believed that water would be found there; all the same, it was not regarded as theoretically impossible. Moreover, the lower gravity would mean that the Martian atmosphere would fall off in density, with increasing height, much more gradually than on the Earth. At 18 miles up, the pressures would be the same, and at greater altitudes the atmosphere of Mars would actually be the denser of the two.

Then, however, new techniques came into play, and there were indications that the atmospheric pressure was much lower than 80 millibars. The Mariner findings were conclusive. On Mars, the pressure of the atmosphere was proved to be below 10 millibars everywhere, and when the Vikings landed, in 1976, they reported pressures of below 7 millibars. The pre-1960 estimates had been wrong by a factor of more than ten.

Estimates of the atmospheric composition had been even worse. Pre-Mariner, the only possible method was to use the spectroscope, which splits up light. Mars shines by reflecting the rays of the Sun, and basically its spectrum is simply a much enfeebled version of the solar spectrum; but a ray of light reaching us from the planet has passed through the Martian atmosphere twice—once on its way from the Sun to Mars, and once on its way back from Mars to Earth. Oxygen and water vapour, if present, would be expected to leave definite imprints upon the spectrum we see, and this also applies to other gases, though nitrogen—which makes up about 78 per cent. of the Earth's air—is very shy about revealing itself in the visible spectrum under such conditions.

The first serious measurements were encouraging. In 1867 Jules Janssen, who founded the Meudon Observatory (the square outside the main entrance is still known as the Place Janssen) took his instruments up to the top of Mount Etna, 9800 feet above sea-level, to study the spectrum of Mars without the handicap of having to look through the denser lower parts of our own atmosphere, which complicates matters because it contains a great deal of both oxygen and water vapour. Janssen's results seemed to indicate that the Martian atmosphere was fairly rich in water vapour, at least. Studies by the great English spectroscopic pioneer, Sir William Huggins, and by Hermann Vogel in Germany provided confirmation—or so it was thought.

The essence of Janssen's method was to compare the spectrum of Mars with that of the Moon. Both shine by reflected sunlight, but the Moon has no atmosphere at all, so that any differences in the two spectra would presumably be due to gases around Mars. Janssen's reasoning was perfectly sound, but his equipment was very crude judged by modern standards, and with hindsight we have to admit that his results were illusory.

The next major step was taken by W. S. Adams and T. Dunham in 1933, at the Mount Wilson Observatory in California. Their instruments were far more sensitive than Janssen's, and they followed a different line of attack. The positions of the lines in a spectrum will be affected by the motion of the body sending out the light; this is the well-known Doppler Effect. If the light-emitting body is approaching, the spectral lines will be shifted over toward the short-wave or 'blue' end of the spectrum; if the body is receding, the shift will be to the 'red' end. When Mars is coming toward us, therefore, the lines due to (say) oxygen in the Martian atmosphere will be blue-shifted, whereas the lines due to the oxygen in our own air will be unaffected. In this way it ought, theoretically, to be possible to disentangle the two sets of lines, and so to work out how much oxygen there is around Mars.

The results were interesting. Adams and Dunham failed to find any oxygen at all, and wrote that "the amount of oxygen in the atmosphere of Mars is probably less than 1/10 of 1 per cent. of that in the Earth's atmosphere over equal areas of surface". Later investigators were equally unsuccessful. In 1947 G. P. Kuiper, also in the United States, announced that he had detected traces of carbon dioxide; in 1963 Dollfus, working at the Jungfraujoch in the Swiss Alps, claimed that there was a certain amount of water vapour, but it was tacitly assumed that the bulk of the Martian atmosphere must be made up of nitrogen. This would have been no surprise. By volume, our own air is made up of about 78% nitrogen, 21% oxygen, and 1% of other gases, mainly argon. Carbon dioxide accounts for a mere 0·03%.

Historically, I must not omit to mention a theory proposed in 1960 by three American astronomers: C. C. Kiess, S. Karrer and H. K. Kiess. They believed that the atmosphere of Mars contained poisonous oxides of nitrogen, and that the polar caps were due to deposits of solid nitrogen tetroxide, while the reddish colour of the planet was caused by nitrogen peroxide. Not many people agreed with this idea, and if it had been correct our hopes of establishing colonies on Mars would have been permanently dashed. Luckily, we now know it to be wrong.

For that matter, the best 'official' analysis of the Martian atmosphere in pre-Mariner days was as wide of the mark as it could possibly be. De Vaucouleurs summed matters up when he listed the probable composition as 98·5% nitrogen, 1·2% argon, 0·25% carbon dioxide and less than 0·1% oxygen. In fact, Viking has told us that carbon dioxide makes up 95% of the atmosphere, with only 1 to 2% nitrogen.

One mystery which remains is that of the so-called Violet Layer or 'blue haze'. All I can do is to give the facts, but I do not pretend to understand them, and some unexpected factor may well be involved.

Light may be regarded as a wave-motion, and the colour depends on the wavelength. In the visible band, red has the longest wavelength and violet the shortest, with orange, yellow, green and blue in between; anyone who has seen a rainbow will be familiar with this order. Considering its comparative thinness, the Martian atmosphere is strangely opaque to short wavelengths. Under normal conditions, a photograph of the planet taken in violet or blue light will appear blurred and slightly enlarged, because the rays do not penetrate to the surface at all, and all that we are photographing is the upper part of Mars' atmosphere. Photographs taken in red light slice through the shielding layers and record surface details, such as the dark areas and the polar caps, with no difficulty at all.

So far as the short wavelengths were concerned, it was claimed that something acted as a screen, and that this 'something' was variable. There were times when it cleared away, but it always came back. This might indicate a definite layer of material which was practically opaque to short-wave light, and this became known as the Violet Layer, not because it looked violet—visually it could not be seen at all—but because it blocked out the violet and blue rays.

The Layer was first reported in 1909, by C. Lampland at Flagstaff. Until 1937 it was assumed to be permanent, but in that year E. C. Slipher, one of the most famous of all observers of Mars, found that at intervals the atmosphere clears "sufficiently to permit the short wavelengths to penetrate to the surface below and come out again". Using the equipment at Flagstaff, Slipher took many hundreds of photographs, and maintained that during a 'clearing' of the Layer, photographs taken with blue or violet filters showed almost as much detail as the red and yellow pictures normally do.

Sometimes the clearing is partial; sometimes it affects the whole of Mars. Major clearings were seen in May 1937, July 1939, October 1941, December 1943, and during the oppositions of 1954 and 1965. In the latter year, according to de Vaucouleurs, it lasted from 25 August to 3 September. There was a partial clearing in 1958, and others in 1964 and 1965.

All sorts of theories were proposed to explain the Layer. One, widely favoured, was that it was due to tiny ice-crystals from 0·3 to 0·4 microns in diameter (one micron is equal to 1/10,000 of a centimetre). This was the view held by Kuiper and Hess, both of whom had recorded clearings. Ice crystals, they wrote, could account for the Layer, and a rise in temperature of only a few degrees would be sufficient to make the crystals 'sublime' —that is to say, change directly from the solid to the gaseous state without passing through a liquid stage. This would at least explain the suddenness of the clearings, but there were various major objections. For instance, in 1963 de Vaucouleurs studied Mars in the ultra-violet region of its spectrum. Ice crystals reflect strongly in the ultra-violet, but Mars does not.

It was also pointed out that if the clearings were due to the sudden disappearance of the ice crystals, one would have to suppose that the temperature rose abruptly and uniformly all over Mars, which did not sound reasonable. The same objection was raised against the idea that the crystals were composed of solid carbon dioxide—which would, in any case, make the Layer more opaque than it was actually observed to be.

In 1960 E. J. Öpik revived an idea that the Layer could be due to small particles of carbon black, and he suggested that the absorbing material was made up partly of these carbon black particles and partly of dust. Clearings would occur when the dust settled down during calm periods. Unfortunately for this theory, the major clearing of 1956 took place at the time of a great dust-storm. Still another theory was due to the Czech astronomer F. Link, who believed that the Layer particles were interplanetary dust which had collected in the upper Martian atmosphere. (*En passant*, it is certain that meteors could be seen from Mars, because the atmosphere, thin though it is, is quite sufficient to act as a shield. Meteors which dash into our own air, and burn away by friction against the air-particles, are usually destroyed at a height of more than 40 miles above the ground, and at this level the atmospheric density is considerably less than it is at the surface of Mars.)

One fascinating study was carried out by S. L. Hess in 1941, when there was a clearing which persisted for several days. The Layer, he said, was normally able to protect the surface of Mars from short-wave radiations, just as the ozone in our own stratosphere protects the Earth and makes life here possible (though let me stress that nobody seriously believed the Violet Layer over Mars to be due to ozone). During a clearing, Mars is exposed to the full short-wave bombardment, which might be expected to damage any living organisms. At that time it was still confidently believed that the dark regions such as the Syrtis Major were caused by vegetation, using the term in its widest sense. Hess wrote that when the Layer vanishes, there should be a temporary halt in the seasonal cycle of development—that is to say, the 'wave of darkening' spreading from pole toward equator as the organisms are revived by the moisture wafted from the ice-caps. This was what Hess found in 1941. During the absence of the Violet Layer, the development cycle stopped, to recommence only when the Layer reformed and conditions became normal once more. This appeared to confirm not only the existence of a definite screen, but also the theory that the dark areas were due to something which lives and grows.

All this sounded highly convincing, but, as I have already said, there are now the most serious doubts about the reality of the 'wave of darkening'. If I may be allowed to give a personal view, I do not believe in it, and I never have. (Random alterations in the outlines of the dark regions are different, and probably genuine.) Visual estimates of the colours and intensities of the dark areas are hard to make with real accuracy, and photographs taken from Earth are never sharp enough to be conclusive; no picture taken with an Earth-based telescope can show as much as can be seen by a visual observer equipped with, say, a 15-inch reflector. Therefore, it is unwise to place too much reliance upon observations of this sort, even when made by astronomers so experienced and so skilful as Hess.

There is another possible explanation, which may sound rather unexpected. Can it be that the so-called clearings are due to changing conditions in the atmosphere of the Earth, rather than in that of Mars? In 1956, according to Slipher, the clearing was studied on the same night by observers in Arizona, South Africa and Australia—and the results were different. It could well be that the clearings were not genuinely Martian, and were recorded only on occasions when the seeing in violet light improved sufficiently to allow for the detection of the low contrast between bright and dark areas. Whether this explanation is valid or not remains to be seen. If an alleged clearing takes place during a period when a Viking-type probe is operating from the Martian surface, we may find out. The mystery remains, but certainly the space-probe results have cast grave doubts upon the existence of a definite Violet Layer.

Clouds are common enough, as has long been known, and some of them lie high in the Martian atmosphere. For instance, a large cloud shaped like a W has often been seen over the region of Mars which has been named Tharsis—now known to be the site of the highest of the volcanoes. Localized white clouds may be seen anywhere, and can be very prominent; so too are the sunrise and sunset hazes. There can be little doubt that all these lofty clouds are made up of ice crystals.

One area which sometimes appears brilliant is Hellas, which lies south of the Syrtis Major, and was called Lockyer Land on the old maps. It is almost perfectly circular, and can on occasions appear as bright as the polar cap, though at other oppositions (as in 1975) it is dull. Formerly Hellas was thought to be a high plateau, over which clouds would be expected to form. Nowadays we know it to be a deep basin, so that clouds condense inside it.

The so-called 'yellow clouds' are not true clouds at all, but may be regarded as dust-storms (Fig. 24). They were predicted as long ago as 1809, by Honoré Flaugergues, and at times they can blot out virtually all details of the planet. Records of them go back for many years, and the major storm of 1971–72 was of

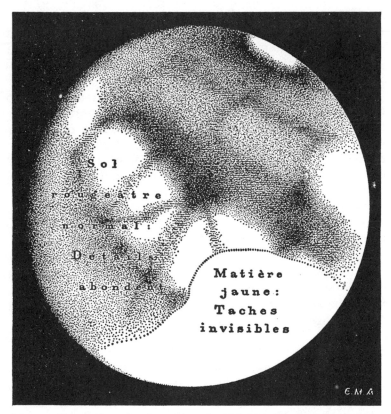

Fig. 24. Vast yellow cloud masking the Trivium Charontis and neighbouring regions, observed by Antoniadi from 23 to 27 August 1929 (8½-in. reflector)

no help to the probes then in the vicinity of Mars—America's Mariner 9 and the Soviet vehicles Mars 2 and Mars 3.

Some of the 'yellow clouds' are transparent and short-lived, but widespread dust-storms may persist for weeks. For some reason, as yet unexplained, they are most frequent when Mars is near perihelion and opposition simultaneously, so that even if they cannot be predicted with any real accuracy we can at least judge when they are likely to occur.

I have tracked the records back as far as I can, and the correlation seems rather too good to be mere coincidence. Thus there were dust-storms in 1909, 1924, 1941, 1956 and 1971–72, all of which were perihelic oppositions; at aphelic oppositions the features have usually been sharp and clear-cut by Martian standards. The most intense storms of all were those of 1909 and 1971–72, each of which was followed by a second storm at the following opposition (1911 and 1973 respectively).

Why should this be so? Certainly the gravitational pull of the Earth cannot have any effect upon conditions in the atmosphere of Mars, and of course we cannot observe the planet every time it reaches perihelion, because it is sometimes on the far side of the Sun. On the other hand I am convinced, from studying the records, that the correlation is real, and the answer must be found in the peculiarities of Martian meteorology.

The storms of 1909, 1911 and 1924 were carefully followed by that great observer of Mars, E. M. Antoniadi, often with the help of the Meudon 33-inch refractor. In August 1924 he reported that "the planet had become covered with yellow clouds, and presented a cream colour similar to that of Jupiter". I was unable to observe that storm personally, as I was then at the early age of one, but I did follow later obscurations, and I particularly recall that of 1956, when Mars was almost as close to us as it can ever come.

Using my 12½-inch reflector, I was able to see the usual surface features throughout August and the first part of September. (Opposition fell on 10 September.) Yet as early as 2 September it became plain that something was amiss. The polar cap was becoming obscure; I could still see it on 8 September, but at that point I decided that my reflector was not large enough to satisfy me, so I took advantage of an invitation to Meudon. From there, on 12 September, I observed Mars under excellent conditions. The disk was practically blank, and remained so for more than a week. The cap reappeared about 19 September, but it was mid-October before the dust-storm finally cleared away.

The next planet-wide storm was that of 1971. It began in

mid-September, and was first photographed by Gregory Roberts, using the 27-inch refractor at Johannesburg. Observers at the Lowell Observatory at Flagstaff reported that it spread out from an initial streak-like core around 1500 miles long, and expanded until by 27 September it covered a large area touching the eastern edge of the famous circular feature Hellas. I could still see Hellas at that time, but it had lost its whitish hue and was no more than very slightly brighter than its surroundings. On 28 September I recorded well-marked dark features in their familiar guise; then came two cloudy nights from my Selsey observatory, and when I could observe again, on 1 October, there was a striking transformation. All I could see were vague indications of the south polar cap and a few indefinite shadings. Within a week I had lost even these, and it was not until the end of December that anything positive could be made out once more—by which time Mars had receded so far from the Earth that details were by no means easy to see even without the added disadvantage of Martian dust.

This, of course, was the period when Mariner 9 was in the neighbourhood of Mars. When it arrived it could do little more than photograph the top of the dust-layer, but the tops of some of the giant volcanoes poked out, which gave a good idea of the altitude of the layer itself.

Because the storm was so violent, I rather expected that another would occur near the next opposition, that of 1973; and it did. For a long period following mid-October the disk, as seen with my $15\frac{1}{2}$-inch reflector and also with the Johannesburg 27-inch refractor, was either blank or virtually so. Yet only a few days before the onset of the storm, the surface features had been absolutely normal in appearance. As so often happens, the planet was covered in a surprisingly short time—less than a week.

Obviously, we have to assume that fast winds occur in the Martian atmosphere. In 1954 de Vaucouleurs made some calculations which indicated that the velocities might exceed 50 m.p.h.; he based this estimate upon the observations of isolated clouds which could be followed as they moved across the Martian disk. In 1967 he stated that "velocities of over 60 m.p.h. have been recorded", though admittedly only for a few hours. Now that the Vikings have told us of gusts reaching at least 35 m.p.h., this sounds reasonable enough even though it is never easy to fix the precise positions of isolated clouds.

The generally-accepted theory was (and is) that the dust-storms are caused by material whipped up from the ochre deserts by surface winds, presumably of the 'whirling' variety. Fast winds must also be produced by the temperature-differences over various parts of the surface, notably near the edges of the polar caps, which would accelerate the process. Fine particles can be held in suspension for long periods, and this was shown by the tremendous outburst of our own Krakatoa in 1883, when most of a volcanic island was blown away and the dust sent upward remained in the stratosphere for a year, producing lovely sunset effects. Moreover, even a thin layer of dust can be remarkably opaque when seen from a distance. On the other hand, even a rapid wind can have relatively little force in that tenuous atmosphere, and it is certainly strange that the dust-storms can be so quick to develop and so extensive.

The only possible loophole is to assume that some dust is sent out from the vents of active volcanoes, but there are very serious objections to this idea.

Some rather special observations have been made from time to time. Projections from the terminator of Mars have been seen often enough, and have been attributed to high-altitude clouds, which is perfectly logical; but there is also the famous case of the 'flare' reported on 8 December 1951 by Tsuneo Saheki, a very experienced Japanese astronomer. The area concerned was that named Tithonius Lacus on the pre-Mariner maps, now known to be the site of the huge canyon which we have christened the Vallis Marineris.

Saheki was using a large telescope under good conditions, and wrote that: "When I first looked at Mars some minutes before 21 hours o minutes, I saw Tithonius Lacus just inside the east limb. Very soon afterwards, a very small and extremely brilliant spot became visible at the east end of this marking. At first I could not believe my eyes, because the appearance was so completely unexpected . . . More careful examination revealed that it was not an illusion, but a true phenomenon on Mars." Subsequently it became brighter than the north polar cap, and then increased in size and faded, vanishing completely in less than an hour.

It was suggested at the time that the flare was due to the eruption of a Martian volcano. This does not seem at all probable, even though Mariner 9 and the Vikings have shown us that the great volcanoes of the Tharsis ridge lie nearby. (Inevitably the Press took up the story; the science reporter on one famous London daily telephoned me to ask what I thought about "the atomic bomb that had gone off on Mars".) On the whole it seems that some kind of cloud phenomenon was involved. Saheki's observation was not unique, but it was, apparently, particularly striking.

To sum up: before the flight of Mariner 4, most astronomers agreed that the atmospheric pressure on Mars was about 85 millibars, that the main constituent was nitrogen, that the Violet Layer provided a screen against short-wave radiations except on the occasions when it temporarily cleared away, and that clouds and dust-storms could be observed frequently. Only the cloud and storm observations remain valid today. The discovery that the atmosphere is much more rarefied than had been expected has affected all our ideas; so let us now turn to the actual surface features which, whatever their nature, can at least be seen with modest telescopes when Mars is well placed in our sky.

Chapter Six

MARS BEFORE MARINER: ICE-CAPS, PLAINS AND DESERTS

In 1666, THE YEAR of the Fire of London, Mars was observed by the Italian astronomer Giovanni Cassini, later to become Director of the Paris Observatory. Using one of the small-aperture, long-focus refractors so common at the time, Cassini saw surface markings on Mars well enough to draw them. He also saw that the polar regions were covered with bright white caps, so that they gave every impression of being snow-covered. Whichever pole happened to be tilted toward the Earth showed this covering, and there was found to be a definite seasonal cycle. The caps were largest in the winter of the hemisphere concerned, smallest in the summer.

A sketch made by Christiaan Huygens in 1656 is said to have indicated the caps, but to be honest I can make little of it, because all it really shows is a dusky band across the centre of the disk which appears to be an optical effect. However, Huygens certainly saw the south polar cap in 1672, and all later observers have recorded them. In fact, they cannot possibly be overlooked when they are at their best. They are not centred on the exact geographical poles, as Giacomo Maraldi pointed out as long ago as 1719. Moreover, they are not alike. The greater range of temperature in the southern hemisphere means that the south cap can become larger than its counterpart, but also that it can become even more reduced in the height of summer. Recent measurements have shown that the remnant of the south cap, when at its smallest, is centred on a point 250 miles from the pole, while with the north cap the difference is only about 40 miles. In midwinter the south cap may extend down as far as latitude −45 degrees.

Early observers, such as Cassini and Maraldi, said little about the nature of the caps, and were content to do no more than follow their changing aspects. William Herschel was more forthcoming, and in 1784 he made the suggestion that the caps were composed of ice or snow. This, of course, was perfectly logical, if only because the Earth has polar caps of such a kind. For many years nobody seriously questioned the snow-and-ice picture, and in what may be termed the final pre-Space Age period of Martian exploration, Antoniadi stated categorically that "this theory of Herschel's is correct".

What was not so clear-cut was the depth of the frozen layer, since it was held that a thick ice-cap would melt and release a great deal of water vapour into the atmosphere—more, indeed, than could be regarded as probable. When a cap shrinks with the arrival of warmer weather it does so quite rapidly, and yet spectroscopes indicated that the Martian atmosphere was virtually bone-dry. A thick cap covering some 4,000,000 square miles (as the southern cap often does) would be expected to make the atmosphere decidedly wet when melting, and this did not seem to fit the facts. As recently as 1954, Gérard de Vaucouleurs wrote that the cap thickness could not be more than a couple of inches, and many astronomers dismissed it as being nothing more than a deposit of hoar-frost, in which case the Martian caps could not be compared with those of our own Antarctica or Greenland. There was also a suspicion that in view of the low atmospheric pressure, the shrinking of a cap could well be due to sublimation (i.e. changing directly from solid into gas) rather than to conventional melting.

Strenuous attempts were made to follow the polar caps through their cycles. It was found that when a cap starts to shrink, its outer rim becomes irregular. With the southern cap, two dark rifts appear, the Rima Australis and the Rima Angusta, and subsequently the cap outline becomes so distorted that a bright white area is left behind, showing up as distinct from the main mass; Schiaparelli, who followed it in 1877, named it the Novissima Thyle. In 1911 the observers of the Mars Section of the British Astronomical Association, at that time directed by Antoniadi, saw it well before it was left behind in the general shrinkage, appearing in the guise of a sparkling white spot well within the main cap; it then showed as a brilliant promontory before the principal cap shrank away and left it isolated. At a later stage in the seasonal cycle, the Novissima Thyle itself usually breaks up into small white dots before vanishing. There are also the so-called "Mountains of Mitchel", first seen by O. M. Mitchel at the Cincinnati Observatory in 1845, which appear as isolated spots at around latitude —73 degrees when the cap is decreasing, but which do not last for more than a week or so. It was generally supposed that these bright spots, like Novissima Thyle, were seen because the snow or ice coating persisted there for some time after it had melted (or sublimed) from the lower-lying areas nearby.

The changes in the northern cap were thought to be similar, and here too there was the regular appearance of a detached white area, Olympia. With each cap, autumn was characterized by the appearance of a whitish overlying haze which sometimes became bright enough to be mistaken for the cap itself, and which hid the growth of the true cap from our inquiring eyes.

Another phenomenon which caused a great deal of argument was the so-called Lowell Band. In his book *Mars and its Canals*, Lowell discussed the theory that the caps might be due to solid carbon dioxide, and wrote: "At pressures of anything like one atmosphere or less, carbon dioxide passes at once from the solid to the gaseous state. Water, on the other hand, lingers on in the intermediate stage of a liquid. Now, as the Martian cap melts it shows surrounded by a deep blue band which accompanies it in its retreat, shrinking to keep pace with the shrinkage of the cap . . . This badge of blue ribbon about the melting cap, therefore, shows conclusively that carbon dioxide is not what we see, and leaves us with the only alternative that we know of: water."

Assuming that the band were real, Lowell's argument was theoretically sound, because moist ground does look darker than dry ground, and carbon dioxide deposits could not possibly show a deep blue band. On the other hand, the reality of the band was questioned. Nobody doubts that the band is visible— I have often observed it myself—but is it due to nothing more significant than contrast between the whiteness of the cap and the much less reflective surface beyond the edge? Antoniadi thought so, and in 1930 wrote that the band was illusory, as he had confirmed by "noting that the band, due to a contrast effect, does not obey the laws of perspective, and that it cannot be photographed". (Yet Antoniadi also wrote that the brightness of the polar cap might well be reduced near the edges by the presence of bushes and grasses, which sounds quaint now!)

Arguments continued over the years, some of them quite

heated. Generally, it was felt that for once Lowell was probably right and Antoniadi wrong. In 1939 de Vaucouleurs made a long series of observations at Le Houga, and decided that even when all possible contrast effects had been taken into account the band was still too dark to be anything but genuine. Later, G. P. Kuiper wrote that "I observed the band with the 82-inch reflector at McDonald Observatory, Texas, under excellent conditions in April 1950, and found it black . . . The rim is unquestionably real; its width is not constant, and its boundary is irregular."

Even temporary lakes or marshes at the boundary of the melting polar cap were thought to be possible, though Audouin Dollfus tended to regard the band as merely part of the famous 'wave of darkening' which was then accepted as a seasonal phenomenon. But—were the polar caps made up of ice and snow at all?

Kuiper, writing in 1949, was in no doubt whatsoever. "The Mars polar caps are not composed of carbon dioxide, and are almost certainly composed of H_2O frost at low temperature, much below $O°C$." The idea of solid carbon dioxide caps had been proposed half a century earlier by two British astronomers, A. C. Ranyard and Johnstone Stoney, but had never become popular, and by the 1950s it had been more or less relegated to the limbo of forgotten things. Neither was there any marked support for the unattractive Kiess theory of solid nitrogen tetroxide. The whole question seemed to be more or less decided.

It is strange how ideas change. The early Mariner results—those of 1965 and 1969—caused a complete swing of the pendulum. Gone were the water-ice or snow caps; instead, carbon dioxide became the fashion, and this view persisted even after the flight of Mariner 9 in 1971–72. It was only with the Vikings that the pendulum swung back once more. Nowadays, it is thought that the residual caps are of ordinary ice, perhaps half a mile thick, and we know that in the northern hemisphere the atmosphere above latitude +60 degrees was saturated with water vapour at the time when the two Viking Landers were operating. Yet there is some solid carbon dioxide too, making up a thin surface layer. As so often happens with two opposing theories, each has turned out to have some truth in it.

85

Next, there was the all-important problem of the nature of the surface away from the polar caps. There was little argument about the ochre tracts, which were generally termed deserts (a term due, as noted, to W. H. Pickering in 1886). Not that there were any serious suggestions that a Martian desert might be a kind of Sahara, with oases, palm-trees, and camels wandering about to admire the view—even though Lowell did compare the colour of the Martian tracts with that of the Painted Desert of Arizona, and the similarity is undoubtedly striking. Subsequently, it was suggested that the oxygen originally present in the Martian atmosphere had combined with the surface rocks, producing a layer of iron oxide—in other words, rust. As Rupert Wildt pointed out in 1934, this would explain both the ochre colour and the present scarcity of free oxygen in Mars' atmosphere.

Simple telescopic observations could shed little further light on the problem, but during the oppositions of the 1920s the question was studied by Bernard Lyot, one of the greatest planetary observers of the present century. Lyot's method was to analyse the light from Mars and compare it with moonlight. He also studied mixtures of grey, brownish and bluish volcanic ash, from which he came to the conclusion that both the Moon and Mars were coated with a layer which was, essentially, volcanic ash. Later on, work by Kuiper in America, Dollfus in France, and others led to the general acceptance of coloured minerals such as felsite or limonite.* This, at least, has been confirmed by the space-probes, so that for once we have a cherished theory which has not been thrown overboard.

Where everyone went complctcly wrong was in the assumption that Mars must have a surface which was no more than gently undulating. I cannot resist yet another quote from de Vaucouleurs, this time dating from 1950: "If there are any mountains on Mars they can scarcely exceed five or six thousand feet in height, and must be more like ancient plateaux than well-marked chains of massive peaks in sharp relief." This does not fit in at all well with the towering volcanoes shown by

---

* For the benefit of those interested in chemical matters, felsite is a rock formed of orthoclase (aluminium and potassium silicate) with quartz grains in occlusion, while limonite is a sedimentary deposit of hydrated iron oxide with the formula Fe_2O_3. $3H_2O$.

Mariner 9 and the Vikings. Another mistake was in supposing the dark regions to be old, depressed sea-beds, because some of them, notably the Syrtis Major, have been found to be lofty.

So far as the dark areas themselves were concerned, the first challenge to the original ocean theory seems to have come from Schiaparelli as long ago as 1863, when he pointed out that the dark regions did not reflect the Sun's image as sheets of water might have been expected to do. (Much later on, V. Fesenkov, a Russian astronomer who specialized in planetary research, calculated that any water surface with a diameter greater than 300 yards ought to betray itself in this way.) As time went by the oceanic theory was abandoned, and most people accepted Liais' suggestion that the dark areas were tracts of vegetation. Pickering's discovery of detail in the dark regions as well as the ochre deserts seemed to provide confirmation. Other observations also fitted in—notably the famous (or, in my view, notorious) wave of darkening, plus undeniable modifications in the outlines of some of the dark areas.

One area which is variable is the patch which Schiaparelli called the Solis Lacus or Lake of the Sun, an elliptical area some 500 miles long and 300 miles wide, with its longer axis lying east-west. It was thus drawn by Maraldi in 1704, and so far as we can tell from the incomplete records it stayed the same until 1926, when the longer axis was found to lie north-south. Later in that year Antoniadi, using the Meudon refractor, drew it as three separate patches, the central one divided from its companions by a dusky 'bridge'. By 1930 all was normal once more, with the longer axis back in its old east-west direction. In 1939 there were fresh changes, and at one time it was seen that the Solis Lacus was made up of a number of small dark spots contained in a generally dusky area. I have been watching the region for the past thirty years, and there seems little doubt that the changes are real. Neither is the Solis Lacus unique in showing variations.

Then, too, there are the reported changes in colour. Quite apart from the 'wave of darkening', it has been claimed that the dark areas are subject to seasonal alterations of hue (Fig. 25). Agreement between different observers has never been complete; thus Antoniadi maintained that the Syrtis Major was bluish-green in winter and brownish in early

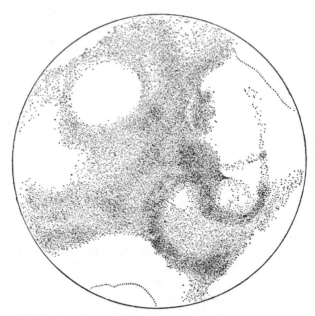

Fig. 25. General aspect of the Syrtis Major region, 13
November 1911: Antoniadi (33-in. refractor)

summer, while in 1964–65 C. F. Capen, at the Lowell Observatory, wrote that "the Syrtis Major was changing from a
blue-green to a green-blue hue . . . The Mare Acidalium
changed from its winter shades of variegated grey and brown
to its spring coloration of dark grey and blue-grey shades with
grey-green oases . . . In late spring and early summer the
Acidalium . . . became a very dark grey general shade with a
black-green central area and large grey-green oases". (Of
course, the term 'oasis' is conventional only, and in no way
marks a return to Lowell's Martians.)

Obviously it is hopeless to rely upon one's eyes for precise
estimates of colour when the intensity is so low. I can make no
useful contribution, because I admit that I have never been
able to see any real coloration in any of the dark regions, but
I do not think that my eyes are particularly sensitive.

An ingenious argument in favour of the vegetation theory was
proposed by E. J. Öpik in 1950. I give it here because it is a
classic example of a theory which appeared too convincing to be

anything but true, and which has nevertheless been shown to be absolutely wrong. Öpik pointed out that windspeeds on Mars are appreciable, and that there are major dust-storms. If, therefore, the dark areas were not due to something which could grow and push the dust aside, it should take only a few centuries for them to become completely covered, giving Mars a monotonous, uniform hue. In pre-Mariner days I think that most people regarded this as conclusive evidence that the dark regions were due to organic material of some kind.

Final proof might, it was thought, come from spectroscopic research. If organic matter could be detected in the spectra of the dark patches, the problem would be solved, and therefore efforts were made to detect chlorophyll, the green colouring matter of so many Earth plants. The results were negative. Chlorophyll looks green because the green light is not absorbed, but is reflected back again; it also reflects infra-red wavelengths, which cannot be seen visually, but which may be recorded on infra-red photographs. It was reasoned that if the Martian areas contained chlorophyll, they should appear bright when photographed in infra-red. Disappointingly, they still looked dark.

On the other hand, G. A. Tikhoff in the U.S.S.R.—founder of what has become known, perhaps futuristically, as the science of 'astrobotany'—pointed out in 1960 that not all forms of vegetation could be expected to show evidence of chlorophyll, and that the spectra of the Martian dark areas could reasonably be compared with the spectra of plants living in very cold regions of the Earth, such as the Siberian tundra. After a long series of experiments he decided that while the detection of chlorophyll would admittedly have proved the existence of Martian organisms, the absence of chlorophyll would not necessarily disprove it. For a while, in 1939, it was thought that spectra taken by W. M. Sinton in America really did indicate traces of organic matter—but then, alas, it was found that the observations had been misinterpreted, and the whole question was thrown wide open again. Therefore, it was pointless to speculate about details of the Martian organisms, since it was by no means certain that they existed at all. Lichens were suggested; so were leafed planets of unknown variety, but all that could really be said was that any Martian life must be of very lowly type.

Various theories were proposed in efforts to explain the dark areas without involving life of any kind. One of these was due to Svante Arrhenius, Swedish winner of a Nobel Prize, in 1912. Arrhenius supposed the dark regions to be overlaid with hygroscopic salts, i.e. salts which can absorb moisture, so that they would darken as soon as they picked up the water-vapour wafted from the melting polar caps. The idea was more or less forgotten, revived in 1947 by A. Dauvillier in France, and then rejected again, mainly because it seemed that there was not enough moisture in the Martian atmosphere to make the process even remotely possible.

Next came a proposal of entirely different kind, due to D. B. McLaughlin of the University of Michigan. In 1954 McLaughlin published a paper in which he claimed that the dark areas were nothing more nor less than volcanic ash, ejected from active volcanoes and distributed by the winds. After discussing the air circulation of Mars as compared with that of the Earth, with particular reference to the Martian equivalents of our own trade winds, he went on as follows: "The sharpish ends of features such as the Syrtis Major point into the wind, and are essentially point sources of some dark material that is carried from these points, fanning out because of variable wind direction. If we restrict ourselves to natural phenomena of which we have experience on Earth, the point sources can have but a single interpretation: they are volcanoes whose ash is carried by the winds and deposited in the pattern we see."

The irregular changes were attributed to outbursts of exceptional violence, and McLaughlin considered that the eruption which caused the change in form of the Solis Lacus in 1926 was as great or greater than the disastrous explosion of Krakatoa in 1883. The whole idea was certainly ingenious. On the other hand, there was general scepticism about the prospect of active volcanoes on a world such as Mars.

Moreover, during the early 1960s there began to be doubts as to whether the dark areas were low-lying after all. Radar studies indicated that some of them might be plateaux. In fact, there was no real certainty about anything; Mars hid its secrets well, and only space-probes could solve them.

I think it is fair to say that if the average astronomer had been questioned in, say, 1960 about the chance of finding life on Mars, he would have replied "Well—probably; but nothing at all advanced". Öpik's argument was most persuasive, and so the findings of the first successful probe, Mariner 4 of 1965, came as a distinct shock. We now know that the edges of the dark areas are not nearly so sharply-defined as they seem when viewed from Earth; there is no basic difference in the character of the terrain between the dark regions and the ochre tracts; there is not even a systematic difference in level, since some of the dark areas are depressed basins while others are high. We are dealing with differences in surface colour only, which was the very last thing which had been expected. All in all, the dark patches seem to be much less significant than they look.

We have come almost to the end of the 'pre-Mariner' summary. But before turning to the probes themselves, it will, I feel, be useful to present something in the nature of a guided tour of Mars as it appears through telescopes set up on the surface of our own world.

Chapter Seven

MARS AS SEEN FROM EARTH

ANYONE LOOKING AT MARS through a telescope for the first time is apt to be disappointed, because generally speaking there is little to be seen apart from a reddish disk, a few dark features, and whiteness over whichever pole happens to be turned toward us. Small instruments will show nothing more, even when Mars is close to opposition. And yet it now seems that on two occasions before the Mariner flights, at least some of the craters were discovered.

The evidence comes from J. E. Mellish, a very experienced and skilful observer, who was able to make use of the world's largest refractor—the 40-inch at the Yerkes Observatory in America. In 1915 Mellish was looking at the planet, and wrote: "There is something wonderful about Mars. It is not flat, but has many craters and cracks. I saw a lot of the craters and mountains this morning with the 40-inch, and could hardly believe my eyes. This was after sunrise, and Mars was high in a splendid sky. I used a magnification of 750."

Subsequently Mellish found that Edward Emerson Barnard, also of the United States, had seen craters on Mars in 1892–93 with the 36-inch refractor at the Lick Observatory, but had not published his results because he thought that people would disbelieve him and make 'fun of it'. Barnard, however, was renowned for his keen sight, and his testimony, coupled with Mellish's, is apparently conclusive. Unfortunately the drawings cannot be produced; those made by Mellish were accidentally burned, while nobody knows what happened to Barnard's. Possibly copies of them will be found one day. But, needless to say, the Martian craters are absolutely inaccessible to ordinary observers using ordinary telescopes, and it is hardly surprising that our ideas in the pre-Mariner era were often so wide of the mark.

In the chart given on page 94 (Fig. 26), drawn from my own observations, I have put in only those features which I am confident of having seen myself with the telescopes in my own observatory, the largest of which is a 15-inch reflector. I make no claim to be lynx-eyed, and I am by no means a good draughtsman, so that no doubt many people would be able to see more than I have done under similar conditions. I feel that there is nothing in the map which is outside the range of a good observer with a 12-inch telescope, but it may help when we come to discuss the various probe results; for instance, if referring to a landing in Chryse, it is a good idea to know just where Chryse is.

Earlier on, I mentioned the official decision to publish all planetary drawings and maps with north at the top. This is apt to confuse the observer who has a conventional astronomical telescope, which gives an inverted image. In the true British tradition of compromise, it therefore seemed to me to be best to print the 'observer's map' sideways and to indicate the various features by figures, so that it can be used either way up.

Next there is the question of nomenclature. Schiaparelli's system, extended and modified by Antoniadi and others, was accepted up to the Mariner era, but in some ways it has now become obsolete. The names had been drawn mainly from mythology, geography or pure astronomy; for instance there were famous dark areas such as the Syrtis Major (a gulf in Libya), the Mare Tyrrhenum (Tyrrhenian Sea), the Mare Acidalium (really the 'Sea of Venus') and so on. Solis Lacus was the 'Lake of the Sun', Lunæ Lacus the 'Lake of the Moon', and Pavonis Lacus the 'Peacock Lake', named after the southern constellation of Pavo, the Peacock. Other names, such as Utopia, speak for themselves, while Hellas is the proper name for Greece.

Unfortunately, we now know that the designations are inappropriate. The Mare Tyrrhenum is not a sea, Solis Lacus is not a lake, and Pavonis Lacus is one of the highest of all the Martian volcanoes. Following the Mariner 9 revelations some new terms were introduced, as follows:

Catena: a chain of craters. Example: Ganges Catena.
Chasma: a canyon, or steep-sided depression. Example: Juventæ Chasma.
Dorsum: a ridge, made up of irregular, elongated prominences. Example: Argyre Dorsum.

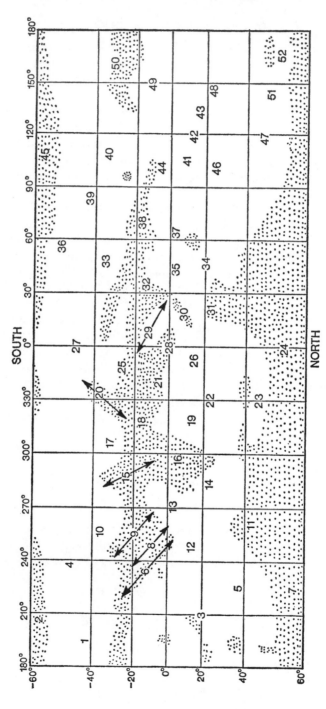

Fig. 26. Outline map of Mars, from observations by Patrick Moore, 1971–76 (15-in. reflector)

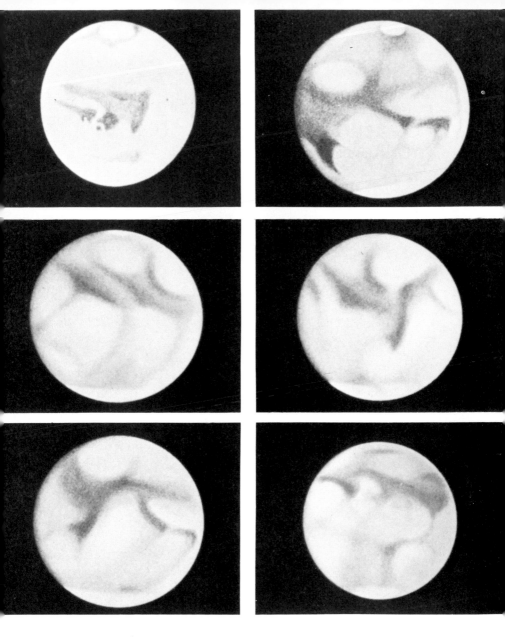

I. *Mars seen from Earth*

1. 28 Aug. 1973, 02.35. 27in. refractor, × 500. Long. of c. meridian 019°. Patrick Moore.
2. 5 Oct. 1973. 22.30. 10in. reflector× 380. Long. 323°. Paul Doherty.
3. 7 Dec. 1975. 21.55. 10in. reflector× 300. Long. 244°. Paul Doherty.
4. 7 Dec. 1975. 23.55. 10in. reflector× 300. Long. 274°. Paul Doherty.
5. 8 Dec. 1975. 02. 05. 10in. reflector× 300. Long. 305°. Paul Doherty.
6. 13 Dec. 1975. 21.30. 10in. reflector× 350. Long. 186°. Paul Doherty.

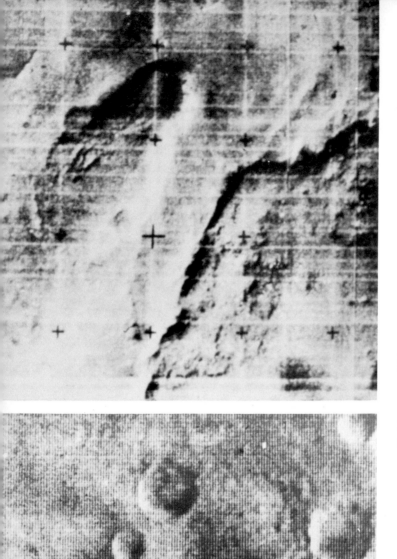

II. *Mariner 4 and Mars-5*

(*left*) Mars from Mariner 4; the Atlantis area, photographed on 14 July 1965 just before the Mariner made its closest approach to the planet. The slant range of the photograph is 7800 miles. Area covered: E-W 170 miles, N-S 150 miles; position lat. 31 °S., long. 197 °E. The Sun was 47° from the zenith. This was the 11th of the Mariner 4 pictures, and probably the best. (*right*) One of the Mars pictures taken by the Russian probe Mars-5, in March 1974. Some detail is shown, but the photograph cannot compare with those of the later Mariners or the Vikings.

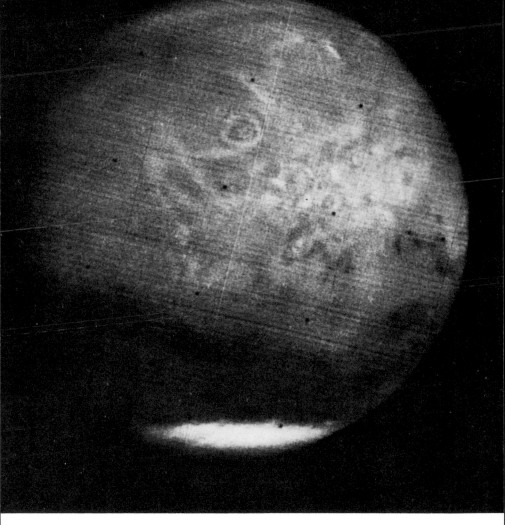

III. *Mars from Mariner 7*

This 'far encounter' picture was taken on 4 August 1969, when the Mariner was 281,000 miles from Mars; the time was 11.14 G.M.T. North is at the top; the longitude of the central meridian is 233°.3. The picture was the 74th in the Mariner 7 sequence. The most prominent feature is the bright Olympus Mons (then still called Nix Olympica) – now known to be a towering volcano. Complex bright streaks are visible nearby in the Tharsis region, and at the right edge there is a dark feature which was in the approximate position of the old 'canal' Agathodæmon. The dark diffuse area to the lower left is the Mare Sirenum. Note also the linear cleft in the south polar cap.

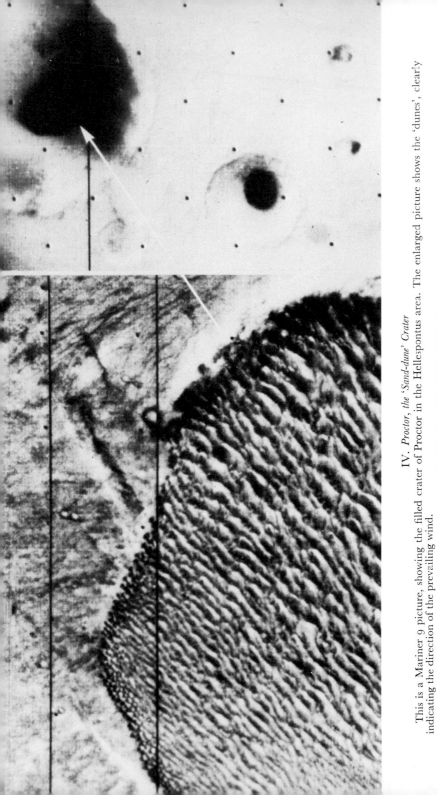

IV. *Proctor, the 'Sand-dune' Crater*

This is a Mariner 9 picture, showing the filled crater of Proctor in the Hellespontus area. The enlarged picture shows the 'dunes', clearly indicating the direction of the prevailing wind.

V. *(below) Part of the 'Valles Marineris'*

The famous Mariner 9 picture of the giant canyon in the Coprates – Tithonius area, now called Valles Marineris. It is undoubtedly part of the drainage system associated with the Tharsis volcanoes, and gives every impression of having been cut by water; note the features which look very like tributaries. A long crater-chain is also very much in evidence.

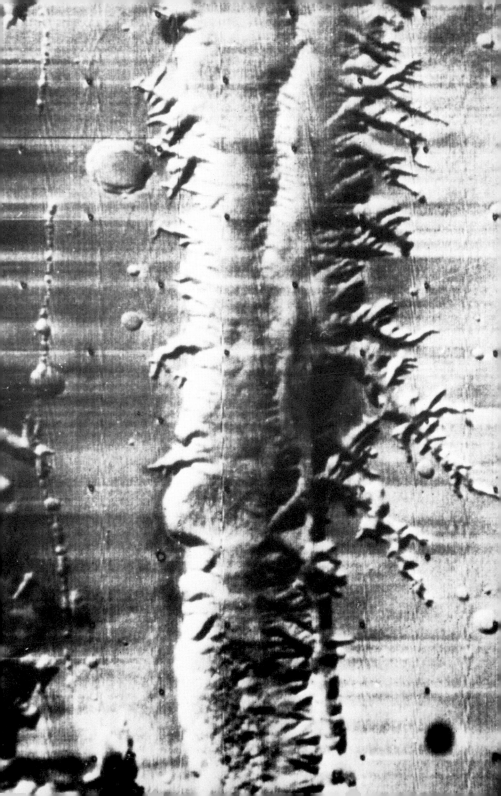

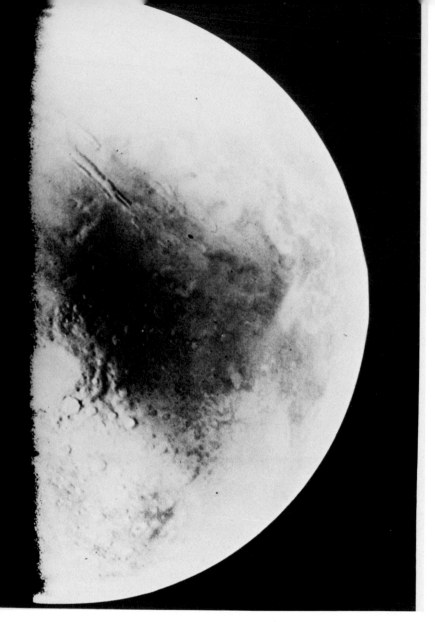

VI. *Mars from Viking 1*

This picture was taken on 18 June 1976, as Viking approached Mars and well before it entered an orbit round the planet. Just below the centre of the picture, and close to the morning terminator, is the basin of Argyre. North of it is the Valles Marineris. The bright area south of Argyre is partly covered by the white south polar deposit. The region to the top of the picture is the Tharsis volcanic area, appearing bright because of clouds. It is notable that the dark regions do not have sharp boundaries – contrary to the overall impression gained from using Earth-based telescopes to observe Mars.

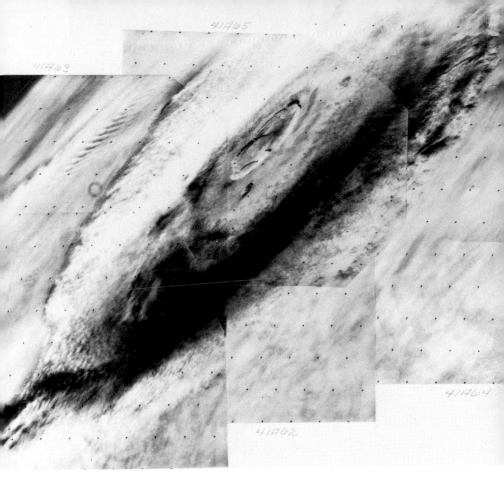

VII. *Olympus Mons*

The great Martian volcano – the highest and most massive volcano known anywhere in the Solar System – was photographed from the Viking 1 orbiter on 31 July 1976, from a distance of 5000 miles. The volcano has an altitude of 15 miles. It is shown here in mid-morning, wreathed in clouds which extend up the flanks to an altitude of about 12 miles. The multi-ringed summit caldera, 50 miles across, pushes up into the Martian stratosphere, and is cloud-free. The cloud cover is most intense on the far western side of the mountain, and a well-defined wave cloud train extends several hundreds of miles beyond the volcano (upper left). The planet's limb is seen to the upper left corner. Extensive stratified hazes are also visible. The clouds are thought to be composed principally of water ice, condensed from the atmosphere as it cools while moving up the slopes of the volcano. The base diameter of Olympus Mons is about 375 miles – approximately the same as the distance between London and Edinburgh.

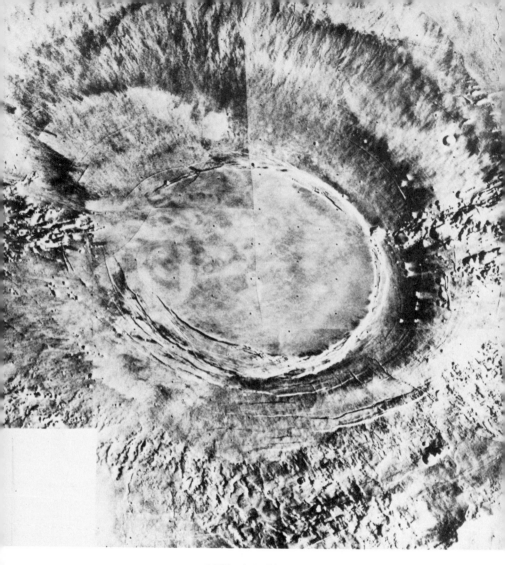

VIII. *Arsia Mons*

In every way comparable with Olympus Mons is the volcano now known as Arsia Mons (its old name was Nodus Gordii). It has an exceptionally large summit caldera, 62 miles in diameter, excellently shown on this Viking Orbiter 1 picture. Like Olympus Mons and some of the other Tharsis volcanoes, Arsia Mons is so lofty that its summit often protrudes above the dust-storms – a fact noted by Schiaparelli long ago, though naturally he had no idea that the features were massive volcanoes. On earlier Mariner 9 pictures Arsia Mons was referred to as 'South Spot'.

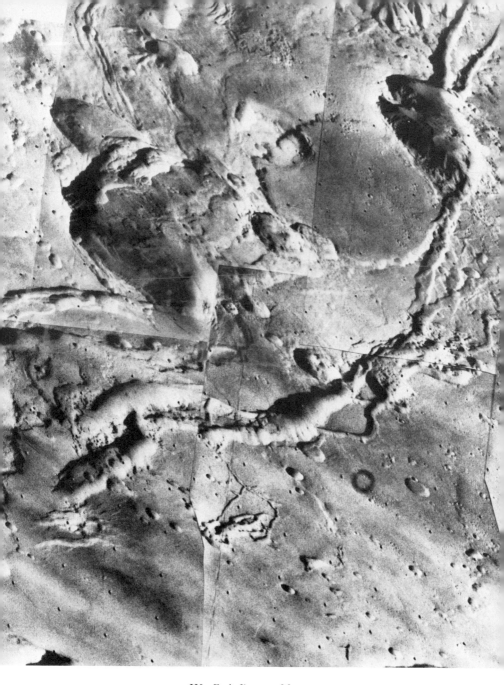

IX. *Fault Zones on Mars*

During surveys of possible landing sites for Viking 2, this picture of an area two degrees south of the Martian equator was taken from Viking 1 on 8 July 1976. Fault zones breaking the crust are clearly shown; the fault valleys are widened by mass wasting and collapse. (Mass wasting is the down-slope movement of rocks under the influence of gravity.)

X. (*top*) First view of Mars from the Viking 1 lander. It was taken a few minutes after the touch-down on 20 July 1976. The large rock to the upper left is about 4 inches across. Many other rocks are shown, together with fine-grained material.
(*above*) Part of the polar cap, photographed from the Viking 1 orbiter. The dark streaks represent surface which is not covered by the white deposit.

XI. (*top*) The 11-mile crater Yuty, in Chryse, photographed from Viking 1 during orbit. Note the terraced walls and the extremely massive central peak, which is crowned by pits and which gives every indication of being a volcanic structure.

(*bottom*) Argyre Planitia, photographed on 11 July 1976 by Viking Orbiter 1. This is an oblique view across Argyre toward the horizon, some 12,000 miles away. The atmosphere was clear, and the horizon brightness due to thin haze; above the horizon are detached layers of haze 15 to 25 miles high.

XII. *The Surface of Mars*

(*top*) The Chryse scene, from Lander 1, photographed on 3 August 1976. The local time is 7.30 a.m. – early morning on Mars. The large rock to the left measures 10ft. by 3ft.; had the Lander come down on top of it, the results would have been disastrous. There are many rocks, some of them sharp, together with what may be termed sand-dune effects. Part of the 'meteorological arm' of the Lander is also seen. (*bottom*) Utopia, from the Lander of Viking 2; an 85-degree panorama sweeping from north at the left to east at the right. It is Martian afternoon; the date is 5 September 1976. Large blocks litter the surface. Some are porous and sponge-like, such as the one to the left edge, which is 1½ to 2 feet across; others are dense and fine-grained, such as the very bright rounded block 1 to 1¼ft. across (lower right). The pebbled surface between the rocks is covered in places by small drifts of very fine material, similar to the drifts seen at the Lander 1 site in Chryse some 4600 miles to the south-west. Some of this material is banked up behind the rocks. On the horizon, to the right, flat-topped ridges or hills are illuminated by the afternoon sun. The apparent slope of the horizon is due to the 8-degree tilt of the spacecraft.

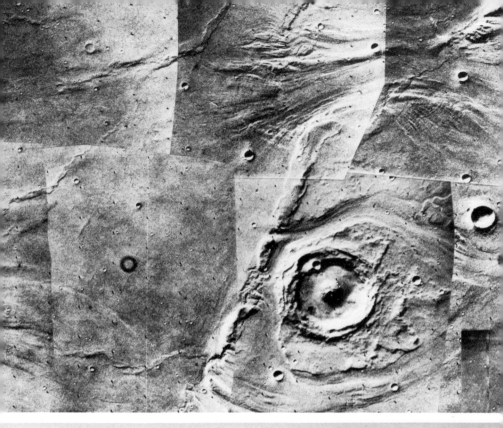

XIII. *Lava-flows and Watercourses*

(*top*) Part of a mosaic of 15 photographs taken around midnight on 9 July 1976 by Viking 1 from a range of 1040 miles. The area is WNW of the original site selected for the landing. Lava-flows are shown, broken by faults which form ridges. Sinuous river channels cross the area. (*bottom*) Utopia, from Viking 2. Winding between the rocks may be seen features which could well represent old stream-beds. The apparent tilt of the horizon is caused by the 8-degree tilt of the space-craft.

XIV. Soil Sampling

(*top*) Collecting samples: Viking 1 Lander, 8 October 1976. The irregularly-shaped rock was pushed several inches by the Lander's collector arm, moving the rock to the left of its original position and leaving it slightly cocked upward. The left-hand picture shows the collector head pushing against the rock - which was named 'Mr. Badger' by the flight controllers. The picture at the right shows the displaced rock and the depression from whence it came. A sample from the site was successfully collected on 11 October.

(*right*) Trench dug by the Viking 1 Lander on Sol 8. The trench is 3 inches wide, 2 inches deep and 6 inches long.

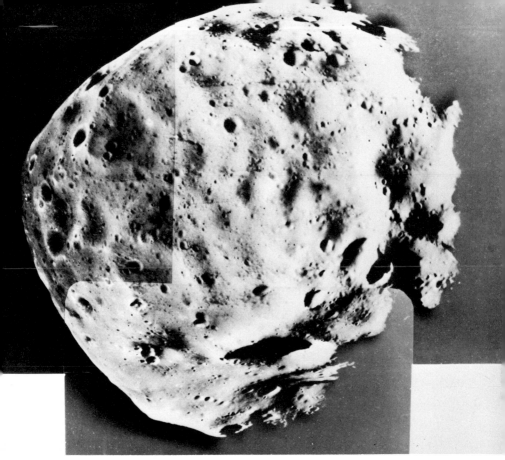

XV. *Phobos*

(*top*) Phobos from Viking Orbiter 1; 22 February 1977. Phobos is 75 per cent illuminated, and is about 13 miles across and 11.8 miles from top to bottom; north is at the top. The south pole is within the crater Hall (3.1 miles in diameter) at the bottom centre, where the pictures of the mosaic overlap. Features as small as 65ft. across can be seen. There are striations, crater-chains, a linear ridge, and hummocks – some of which are over 160ft. in size. A long linear ridge extends from the south pole toward the upper right, and a series of craters runs horizontally across the picture, which is parallel to the orbital plane of Phobos. The mosaic was taken during the Orbiter's 242nd revolution round Mars.

(*left*) Another Viking 1 view of Phobos, taken from a range of 545 miles; the camera was slewed to compensate for the rapidly-changing motion at this range. The smallest visible feature is 130ft. across; Phobos itself is visible for 11 miles from top to bottom and 5.6 miles from left to right. Note the remarkable striations, unknown before Viking 1 took detailed photographs of the satellite.

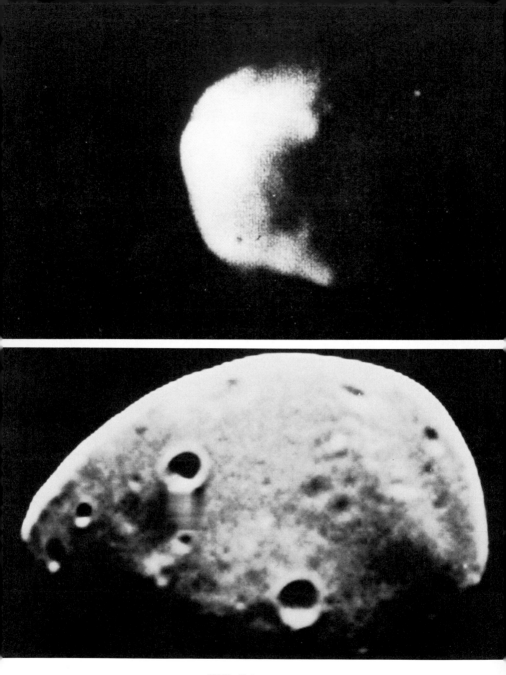

XVI. *Deimos*

(*top*) Preliminary photograph of Deimos from Mariner 9.
(*bottom*) This photograph was taken from the orbiter of Viking 1. The largest craters recorded are Voltaire (diameter 1.2 miles) and Swift (0.6 miles); of the two Swift has the sharper outline. In shape Deimos is roughly ellipsoidal, with principal diameters of $9\frac{1}{2}$, $7\frac{1}{2}$ and 7 miles. The surface is dark, with an average albedo of 6 per cent., though in one patch the reflectivity rises to 8 per cent. Like Phobos, Deimos has synchronous rotation – that is to say, it keeps the same face turned toward Mars all the time.

KEY TO MAP

Fossa: a 'ditch'—a long, narrow, straight or curved valley. Example: Sirenum Fossa.

Labyrinthus: a valley complex. The only example so far named is Noctis Labyrinthus (formerly known as Noctis Lacus).

Mensa: a flat-topped, table-like area with steep slopes to either side. Example: Nilosyrtis Mensa.

Mons: a mountain. Example: Olympus Mons (which is, of course, a volcano). Olympus Mons was formerly called Nix Olympica, the Olympic Snow, and had long been suspected of being lofty. Another famous mountain is Arsia Mons, previously known as Nodus Gordii, the Gordian Knot.

Patera: a shallow, complex crater with scalloped edges. Example: Alba Patera.

Planitia: Smooth, low plain. Examples: Hellas Planitia, Argyre Planitia, Chryse Planitia, Utopia Planitia. Hellas and Argyre were once thought to be plateaux, but are in fact basins; Hellas is the deepest known depression on Mars.

Planum: a plateau, or high plain. Examples: Lunæ Planum, Solis Planum.

Tholus: a hill or an isolated, dome-shaped mountain. Example: Hecates Tholus.

Vallis: a valley—a sinuous channel, often with tributaries. The best example is Vallis Marineris, or Mariner Valley, about which more anon.

Vastitas: Extensive lowland plain. The vast circumpolar northern plain is named Vastitas Borealis; it used to be called the Mare Boreum.

To avoid as much confusion as possible, I have used the new names. A few of the old designations of the albedo features have been retained, notably Margaritifer Sinus (the Gulf of Pearls), Sinus Sabæus (named after the Red Sea) and Sinus Meridiani (the Meridian Bay, which was always used as the zero longitude for Mars—in fact, the Martian 'Greenwich', chosen quite arbitrarily). These three dark markings seem to have nothing to distinguish them from their surroundings, apart from their colour.

In other cases I have thought it best to keep to the older names as given in the official map compiled by the International Astronomical Union committee shortly before the Mariners began to fly. For instance, there are as yet no new names for the ochre tracts which have always been known as Aeria, Phæthontis and so on.

Obviously, the observer has to depend upon the tilt of Mars with respect to the Earth. When the southern hemisphere is favoured, the Syrtis Major is pre-eminent; when the northern hemisphere is presented, pride of place generally goes to Acidalia Planitia (formerly the Mare Acidalium), though the Syrtis Major is conspicuous even then. As we have noted, the southern hemisphere is tipped toward us at perihelic oppositions, which is why it was better-mapped than the northern region until the space-ships took over.

Here, then, is a very brief 'tour' of Mars which may be undertaken by anyone equipped with an adequate telescope of, say, 8 inches aperture.

It is convenient to begin with the *Syrtis Major* (I still have an unscientific wish to call it the Hourglass Sea!) which is V-shaped, and extends from the equator into the northern hemisphere. It is said to show changes, some of which are seasonal while others are unpredictable; there are reports that it is sometimes relatively narrow, sometimes broad. I cannot comment, because I have never been sure of these variations. They are admittedly well-authenticated, though not so easy to explain as they used to be when astronomers still believed the Syrtis to be a vegetation-tract.

To the west is the ochre region of *Aeria*, which merges into *Arabia*. To the east is *Isidis Planitia* (formerly Isidis Regio). Well to the north is *Utopia Planitia*, celebrated as being the landing-site of Viking 2 in September 1976.

Syrtis Major is part of a dark mass which extends for more than half-way round the planet. It adjoins a rather narrow but often very prominent dark region, the *Sinus Sabæus*, which is separated from a similar region, *Pandoræ Fretum*, by the lighter region of *Deucalionis*.

To the south of the Syrtis Major is one of the most famous markings on Mars: *Hellas Planitia*, which is circular and practically featureless. (In the 'canal' days it was recorded as being crossed by two streaks making up an X; they were named the Peneus and the Alpheus, but, alas, they do not exist.) Hellas is very variable in brightness. Sometimes, as in 1967, it can rival the polar cap; at other oppositions, as in 1975, it is hardly identifiable. There is no mystery about these changes. When Hellas is bright, its basin is cloud-filled. Adjoining it, between it and the tract of *Noachis*, is a dark region, *Hellespontus*; it is heavily cratered, though of course nothing of the kind was expected before the probe pictures. *Hadriaca Patera*, to the east, is rather similar. (This was formerly Mare Hadriacum—the Adriatic Sea.)

Extending from the Syrtis Major and the adjoining *Libya*, to the east, are two more dark regions, *Tyrrhena Patera* (once the Mare Tyrrhenum) and *Cimmerium*, separated by the lighter *Hesperia Planum*, which is sometimes—not always—easy to identify. To the north are *Electris* and *Eridania*, two more of the ochre tracts. It was in this area that the capsule of the Russian probe Mars 3 came down in 1971, though without sending back any useful information after its arrival.

Well to the north of the Tyrrhena-Cimmerium streaks is the *Trivium Charontis*, a darkish patch which can be quite prominent at times, and was once thought to be the centre of a system of radiating canals with fascinating names such as Cerberus, Hades, Erebus and Styx. Only the *Cerberus* has any real existence, and it is certainly not a canal.

In this general area are various lightish areas such as *Æthiopis*, *Ætheria*, *Elysium Planitia* and *Isidis Planitia*. On older maps the boundary between Æthiopis and Isidis was marked by one of the most celebrated of the canals, the Nepenthes-Thoth; I have seen a diffuse dusky streak there running in the direction of the wedge-shaped, darkish area of *Casius*.

Now let us go back to the equator. Here we find the Meridian Bay—*Sinus Meridiani*, which has two dark 'forks' pointing northward; it has been called 'Dawes' Forked Bay' and also the Fastigium Aryn, and it marks the zero for Martian longitudes. It extends from the Sinus Sabæus, and under good conditions the forked appearance may be seen with a modest telescope, though it is not always conspicuous and may well be variable in intensity. To its west, and separated from it by a bright region, is the *Margaritifer Sinus*, which is shaped rather like the Syrtis Major, but is much less prominent. Almost due north of it is the *Acidalia Planitia* (Mare Acidalium), with its extension still called the *Niliacus Lacus*, though no doubt the name will be changed before long. Acidalia is the principal dark zone north of the Martian equator, and at aphelic oppositions it tends to dominate the scene whenever it lies on the Earth-turned hemisphere.

Between Acidalia and Margaritifer is *Chryse Planitia*, the first site from which signals from Mars were sent back in July 1976. It merges into the similar region of *Xanthe*, and to the west of Xanthe is the dark *Lunæ Planum* (formerly Lunæ Lacus). To the north-west there are ochre tracts such as *Tempe* and *Arcadia Planitia*. Between these two regions the old maps showed a canal, the Ceraunius; this has gone the way of all canals, but at least it has given its name to a lofty hill, the Ceraunius Tholus, which lies in the volcanic area of *Tharsis*.

It is here that we find the greatest of all the volcanoes, the *Olympus Mons*. Also in Tharsis are *Ascræus Mons, Pavonis Mons* and *Arsia Mons*—née Ascræus Lacus, Pavonis Lacus and Nodus Gordii respectively. All were recorded from Earth long before probes had become even remotely practicable, and at times they have been seen even when the rest of Mars has been veiled in dust, for the excellent reason that they are high enough to poke out above the dusty layers.

The darkish patch between Margaritifer to the one side and Tharsis on the other is *Auroræ Planum*, which extends toward the sometimes-prominent dark region still called the *Tithonius Lacus*. Here Lowell showed a canal, the Coprates; Mariner 9 revealed the immense Vallis Marineris, which is on a giant scale even by Martian standards. Another dark region, the *Mare Erythræum*, lies south of Auroræ; south again is the basin of *Argyre Planitia*, which is of the same type as Hellas, and can be cloud-filled and bright. *Thaumasia* is ochre, and between it and the dark *Phœnicis* is the celebrated *Solis Planum*, which is certainly worth watching. Sometimes it is an easy object, though at other oppositions I have failed to identify it at all even though by all logical standards I ought to have seen it.

There is another prominent dark area in the south-west, *Mare Sirenum*, which has a beak-like extremity; from here Lowell drew a canal, the Araxes, linking Sirenum with Phœnicis, but it too was a mere illusion. *Phæthontis*, yet another ochre region, lies south of Sirenum, with *Memnonia* and *Amazonis Planitia* to its north. Finally, in the north-west adjoining

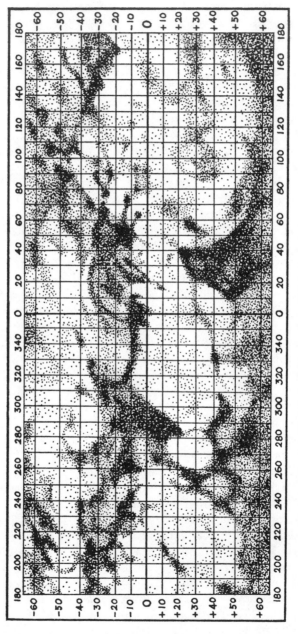

Fig. 27. The I.A.U. Map of Mars—the official chart before the Mariner flights.

Acidalium M. (30°, +45°)
Æolis (215°, −5°)
Aeria (310°, +10°)
Aetheria (230°, +40°)
Aethiopis (230°, +10°)
Amazonis (140°, 0°)
Amenthes (250°, +5°)
Aonius S. (105°, −45°)
Arabia (330°, +20°)
Araxes (115°, −25°)
Arcadia (100°, +45°)
Argyre (25°, −45°)
Arnon (335°, +48°)
Aurorae S. (50°, −15°)
Ausonia (250°, −40°)
Australe M. (40°, −60°)
Baltia (50°, +60°)
Boreum M. (90°, +50°)
Boreosyrtis (290°, +55°)
Candor (75°, +3°)
Casius (260°, +40°)
Cebrenia (210°, +50°)
Cecropia (320°, +60°)
Ceraunius (95°, +20°)
Cerberus (205°, +15°)
Chalce (0°, −50°)
Chersonesus (260°, −50°)
Chronium M. (210°, −58°)
Chryse (30°, +10°)
Chrysokeras (110°, −50°)
Cimmerium M. (220°, −20°)
Claritas (110°, −35°)
Copaïs Palus (286°, +55°)

Coprates (65°, −15°)
Cyclopia (230°, −5°)
Cydonia (0°, +40°)
Deltonon S. (305°, −4°)
Deucalionis R. (340°, −15°)
Deuteronilus (0°, +35°)
Diacria (180°, +50°)
Dioscuria (320°, +50°)
Edom (345°, 0°)
Electris (190°, −45°)
Elysium (210°, +25°)
Eridania (220°, −45°)
Erythraeum M. (40°, −25°)
Eunostos (220°, +22°)
Euphrates (335°, +20°)
Gehon (0°, +15°)
Hadriacum M. (270°, −40°)
Hellas (290°, −40°)
Hellespontica Depressio (340°, −60°)
Hellespontus (325°, −50°)
Hesperia (240°, −20°)
Hiddekel (345°, +15°)
Hyperboreus L. (60°, +75°)
Iapigia (295°, −20°)
Icaria (130°, −40°)
Isidis R. (275°, +20°)
Ismenius L. (330°, +40°)
Jamuna (40°, +10°)
Juventae Fons (63°, −5°)
Lastrygon (200°, 0°)
Lemuria (200°, +70°)
Libya (270°, 0°)

Lunae Palus (65°, +15°)
Margaritifer S. (25°, −10°)
Memnonia (150°, −20°)
Meroe (285°, +35°)
Meridianii S. (0°, −5°)
Moab (350°, +20°)
Moeris L. (270°, +8°)
Nectar (72°, −28°)
Neith R. (270°, +35°)
Nepenthes (260°, +20°)
Nereidum Fr. (55°, −45°)
Niliacus L. (30°, +30°)
Nilokeras (55°, +30°)
Nilosyrtis (290°, +42°)
Nix Olympica (130°, +20°)
Noachis (330°, −45°)
Ogygis R. (65°, −45°)
Olympia (200°, +80°)
Ophir (65°, −10°)
Ortygia (0°, +60°)
Oxia Palus (18°, +8°)
Oxus (10°, +20°)
Panchaia (200°, +60°)
Pandorae Fretum (340°, −25°)
Phaethontis (155°, −50°)
Phison (320°, +20°)
Phlegra (190°, +30°)
Phœnicis L. (110°, −12°)
Phrixi R. (70°, −40°)
Promethei S. (280°, −65°)
Propontis (185°, +45°)
Protei R. (50°, −23°)
Protonilus (325°, +42°)

Pyrrhae R. (38°, −25°)
Sabaeus S. (340°, −8°)
Scandia (150°, +60°)
Serpentis M. (320°, −30°)
Sinai (70°, −20°)
Sirenum M. (155°, −30°)
Sithonius L. (245°, +45°)
Solis L. (90°, −28°)
Styx (200°, +30°)
Syria (100°, −20°)
Syrtis Major (290°, +10°)
Tanais (70°, +50°)
Tempe (70°, +40°)
Thaumasia (85°, −35°)
Thoth (255°, +30°)
Thyle I (180°, −70°)
Thyle II (230°, −70°)
Thymiamata (10°, +10°)
Tithonius L. (85°, −5°)
Tractus Albus (80°, +30°)
Trinacria (268°, −25°)
Trivium Charontis (189°, +20°)
Tyrrhenum M. (255°, −20°)
Uchronia (260°, +70°)
Umbra (290°, +50°)
Utopia (250°, +50°)
Vulcani Pelagus (15°, −35°)
Xanthe (50°, +10°)
Yaonis R. (320°, −40°)
Zephyria (195°, 0°)

Arcadia Planitia, are two darkish patches, *Castorius* and *Propontis*, which I have usually seen without much difficulty when the northern hemisphere has been tilted toward us at a favourable angle.

The popular zones are naturally hard to examine, and there is no point in saying much about them here, but I ought to mention two bright patches in the far south, *Thyle I* and *Thyle II*, as well as the huge northern *Vastitas Borealis* (once Mare Boreum), which extends all round the planet and is covered by the polar cap material during winter.

Again I stress that this is a rough outline guide, in no way a precision chart, but I hope it will be good enough to show the main features which are accessible to the average, well-equipped amateur observer. For the sake of completeness I also give the map which was drawn up under the official auspices of the International Astronomical Union, regarded as the best possible chart in pre-Mariner days, though to my mind it shows far too many streaks which could be classed as canals!

Amateur observers can still make themselves useful, because they can record clouds and other atmospheric phenomena and also note any possible modifications in the outlines of the dark areas. Yet for our real knowledge of Mars we must turn to the space-ships, and begin on 1 November 1962, when Man's pioneer probe was sent on its way to the Red Planet.

Chapter Eight

SPACE-SHIPS TO MARS

THE FIRST MARS-SHIP took off when the Space Age was only a little more than five years old. The opening of the new era may be dated very precisely: to 4 October 1957, when Russia's Sputnik 1 sped round the world sending back 'bleep! bleep!' signals which, to some Western ears, sounded faintly derisory. (An American admiral named Rawson earned a place in history by claiming that the Sputnik was "a hunk of old iron that almost anybody could launch", though at that time the United States space programme was floundering hopelessly.) Probes to the Moon followed in 1959; Yuri Gagarin made his pioneer ascent in April 1961. It was inevitable that vehicles to the planets should follow.

In fact, the Russians made an attempt even before Gagarin's flight. There were only two planets within practicable range: Venus and Mars, and of these Venus had claims to being regarded as the better bet. The revelations about its immensely hostile nature did not come until later, and in 1961 it was still believed that the surface might be reasonably welcoming. On 12 February the Russians dispatched a Venus probe, but after it had receded to less than five million miles all contact with it was lost.

By this time America's programme was well under way, and two Venus probes were launched from Cape Canaveral during the summer of 1962. The first, Mariner 1, failed. The other, Mariner 2, made a successful pass of Venus in mid-December, and sent back the first positive information about that somewhat sinister world. Meanwhile the Russians had turned their attention to Mars.

Nowadays, when flights to the Moon have become history and the launching of a new probe does not even merit a note in the daily papers, it is strange that so many people still have completely wrong ideas about the way in which a space-ship is launched. There is a persistent view that the vehicle has to 'get out of the Earth's gravity'. In fact there is no possibility of anything of the sort, and no need for it. The Earth's gravitational field weakens with increasing distance, but in theory it has no limit. What has to be done is to work up to escape velocity, so that the gravitational tug is inadequate to bring the probe back.

Though the basic notion of space-flight is very old, credit for putting it upon a properly scientific footing must go to a shy, deaf Russian teacher named Konstantin Tsiolkovskii, who published some technical papers at the beginning of the present century which were decades ahead of their time, and of which nobody took the slightest notice (mainly, it is true, because they were in Russian, and came out in an obscure journal). Tsiolkovskii knew that the main problem would be that of fuel, and he suggested using a compound launching vehicle made up of several rockets mounted one on top of the other. At launch, the large lowermost rocket would provide the power; when it had used up all its fuel it would break away and fall back to the ground, leaving Rocket No. 2 to carry on with its own motors. When it, too, had run out of fuel, the uppermost stage would have been put into a path which would take it to its target world.

This is how all lunar and planetary probes have been sent on their way, but the whole procedure is remarkably complicated, and there are grounds for suggesting that the safe landings of the Vikings on Mars in 1976 rank at least equal with the manned lunar landings of the Apollo missions, considered from a purely technical point of view. The main difference between a lunar voyage and a planetary journey is that the planet does not stay obligingly close to us, and of course the distances involved are much greater. Venus is always at least a hundred times as remote as the Moon, and Mars is further away still.

I do not propose to say much about rockets, because it would be too much of a digression. I must, however, stress that a rocket works by what Isaac Newton called the 'principle of reaction'— every action has an equal and opposite reaction. In a rocket, gases are sent out through the exhaust, and these kick against the rocket body and propel it onward. This is why a rocket can function in vacuum; there is no need for any surrounding atmosphere, and in fact air is a positive nuisance, because it sets up friction and this in turn produces heat. A space-rocket takes off slowly, and accelerates to full velocity only when it is safely beyond the dense lower reaches of the Earth's atmospheric mantle. The motors are complex by any standards, and use liquid propellants; but the underlying principle is no different from that of the Guy Fawkes firework rocket.

With unlimited fuel supply, there would be no actual need to work up to escape velocity at all, but practical considerations alter the whole situation. What we cannot do is to wait until a planet (such as Mars) is at its closest to us and then simply fire a rocket across the gap. Quite apart from the numerous other objections, this would mean using power throughout the journey, and no vehicle we can build at the moment could possibly carry enough fuel. We must make use of the Sun's gravitational force, and 'coast' for most of the way.

The Earth moves along at a mean velocity of $18\frac{1}{2}$ miles per second, or around 66,000 m.p.h. If it moved faster, it would follow a different orbit, and would initially swing outward; if it were slowed down by some miraculous means, it would at first swing inward. For a Venus probe, then, the principle is to take the vehicle up in a rocket launcher and then slow it relative to the Earth, so that it enters a transfer orbit and comes within range of Venus. But let us concentrate on Mars, where the probe has to be speeded up.

The diagram (Fig. 28) given here shows the path of Mariner 4, because this was the first Mars probe to be successful. It was launched from Cape Canaveral on 28 November 1964. The

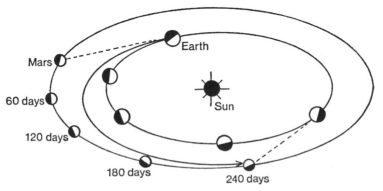

Fig. 28. Flight path of Mariner 4

massive, compound vehicle carrying the Mariner rose majesti-
cally into the air with what seemed to be agonizing slowness; the
lowermost rocket—an Atlas—fired its motors, soon shedding
two of its 'boosters' and using the remaining one. When its
work was done, Atlas broke away and fell back into the sea.
Rocket No. 2, an Agena type, took over at a height of 100
nautical miles and a velocity of 13,000 m.p.h. The Agena
engine was shut down as soon as Mariner 4 had been put into
a closed orbit round the Earth, moving at a brisk 17,500 m.p.h.
After another forty-one minutes the Agena fired again, and
when it finally shut down for good the space-craft was moving at
25,598 m.p.h. in a path which would take it to within striking
distance of Mars. By now it was on its own; Agena, like Atlas,
had broken away and fallen back to Earth.

Basically, no more power was needed for the main journey.
Mariner was in a transfer orbit, and if the calculations were
right the probe would reach the orbit of Mars to rendezvous
with the planet. This duly happened. Within two days the
Mariner was half a million miles from Earth, and it coasted
along until the Mars rendezvous on 14 July 1965 at a mere
6118 miles from the planet. The time elapsing between launch
and rendezvous was 228 days, or 222 sols, and the total distance
covered was approximately 330,000,000 miles. Recalling that
Mars can approach the Earth to within 35,000,000 miles, this
sounds a long way—rather like driving from London to
Bognor Regis by way of Edinburgh—but in terms of fuel it was
the most economical path.

Of course, I have over-simplified matters grossly. A mid-
course correction had to be carried out on 5 December 1964; a
command was sent to the probe, and was duly obeyed. There
was also the question of 'attitude', because if the Mariner
pointed in the wrong direction it would be unable to communicate or to receive proper instructions. The method of achieving
this was most ingenious. The star Canopus was used as a 'lock',
as well as the Sun.

The best way to explain this, I think, is to picture a weight
which is hung from a long cord. It will tend to spin, but a second
cord, at approximately right angles, will steady it. A line of sight
on Canopus was used as Mariner's second 'cord'. Canopus, the
second brightest star in the sky, was eminently suitable; it was

in the right position relative to the Sun, and its brilliancy enabled the Mariner sensors to locate it, though admittedly only after a day of searching around. The method worked splendidly, and even when the Canopus lock was temporarily lost, at the time of the December mid-course correction, the sensor was able to find it again. Power, obtained by utilizing the solar paddles which gave Mariner its characteristic appearance, was no real problem.

Mariner 4 made a single fly-by of Mars. Though its main task was then over, it did not leave the Solar System; it was moving in a stable path round the Sun, so becoming a tiny artificial planet, and there is no reason to suppose that it will not continue in its path indefinitely, though all contact with it was lost when its power eventually gave out.

I have gone into some detail about the Mariner 4 path because it is absolutely typical, but, as we have noted, Mariner was not the first Mars probe—and this brings us back to 1 November 1962, when the Russians made their pioneer effort.

In those early days of space research the Soviet teams had consistent trouble with their long-range communications, and even at the present time (1977) they have had very little luck with Mars. Their first probe, Mars 1, was quite a massive vehicle, weighing almost 2000 pounds, and apparently it was put into the correct orbit, so that all seemed to be well. The path which would take it to Mars was of the same general type as those of the later Mariners, and it carried a variety of instruments, including several cameras.

Like all other probes, too, Mars 1 was designed to carry out studies of the conditions in interplanetary space. Astronomers were—and still are—interested in the so-called solar wind, which is made up of low-energy atomic particles ejected from the Sun in all directions. Also, the magnetic fields in space are of tremendous theoretical importance; cosmic radiation is another field in which probes are invaluable, and, of course, there is the question of meteoritic particles. Not so many decades ago it was still thought possible that any space-ship daring to leave the protective screen of atmosphere round the Earth would be promptly and fatally battered by a concentrated bombardment of meteoroids. Luckily the danger has been found to be negligible, but there are plenty of micro-meteorites, too small to cause any damage to a space-probe but large enough to be recorded. All the vehicles to the planets have sent back details of the number of hits.

By mid-March 1963 Mars 1 was almost 70,000,000 miles from the Earth, and still going well. Then, abruptly, contact with it was lost, and was never regained. What presumably happened is that the 'star lock' failed, so that the probe swung round and was unable to continue sending or receiving messages. In all probability it passed Mars at around 193,000 miles on 19 June 1963, and there is little doubt that it is still orbiting the Sun, but its fate will never be known. By the time it 'went silent', it was so remote that signals from it, moving at the velocity of light, took twelve minutes to reach the Earth.

Chronologically, Mars 1 was the fourth planetary probe. Earlier vehicles had been Russia's Venera 1, which also went out of contact before getting anywhere near its target, and the two American Venus probes, Mariner 1 (which was a complete failure; it went out of control as soon as it had been launched, and had to be destroyed) and Mariner 2, which made a triumphal pass of Venus in December 1962. Following Mars 1, the Russians attempted another Venus shot with Zond 1, and again they failed. In November 1964 the centre of activity, so far as Mars was concerned, had swung back to the United States.

Two Mariners had been prepared: Numbers 3 and 4. They were identical, and each was designed to by-pass Mars and send back data, including pictures. The reason for building two probes instead of only one was mainly as a safeguard in the event of failure. As events proved, this was a wise precaution.

As with all space-craft, the time of launching was important. It is essential to use as little fuel as possible, and the velocity needed to reach Mars is least when the Earth launch and Mars rendezvous occur on opposite sides of the Sun. The 'window', or period of time when a launch is practicable, is limited to a few weeks every two years. Absolutely ideal launch conditions would be when the take-off point, the Sun, and the arrival point are lined up, but this hardly ever happens, because the orbit of Mars is appreciably tilted with respect to that of the Earth. The inclination is only 1·9 degrees, but this is quite enough to make a considerable difference.

The 'window' for 1964 fell in November. At noon on November the Fifth, Mariner 3 was sent up from Cape Canaveral.* Americans do not celebrate Guy Fawkes' Day, but it was certainly an unlucky time for Mariner 3, because although the launch seemed at first to be successful the planners soon realized that the flight was doomed. During the first rush through the Earth's dense lower air, the delicate space-craft itself is protected by a shield, which is jettisoned as soon as the main atmosphere has been left behind. With Mariner 3, the shield stuck obstinately in position, and the dead-weight meant that the velocity was reduced, so that there could be no hope of the probe reaching Mars. Five and a half minutes after the space-craft had been separated from the Agena stage of the launcher, it was ordered to extend its solar panels, which would enable it to use the Sun's energy to provide power for its various items of equipment. Alas, the solar panels failed too, and without them there was no power. Frantically the planners suspended all scientific operations in the Mariner, kept on trying to persuade the panels to extend, and then decided to fire the motor of the space-craft itself in an attempt to jerk the awkward shield free. Before they could do so, the battery ran down. To all intents and purposes Mariner 3 was dead; it had remained in contact for only 8 hours 34 minutes. Silent and untrackable, it entered a path round the Sun, and no doubt it is orbiting even now.

The space-planners were disappointed, but not dismayed. They managed to identify the cause of the problem, and the back-up probe was modified accordingly. By 28 November, Mariner 4 stood on its launching pad; the usual procedure was followed, and this time there was no mishap. The shield dropped away, the solar panels worked perfectly, and the first American messenger to Mars was well and truly on its way. On the following 14 July the rendezvous manœuvre was begun, and shortly after midnight G.M.T. on 15 July Mariner made its closest approach to Mars. For the record, the exact time was 01 hours 0 minutes 57 seconds; the distance from the Martian surface was 6118 miles. By then the picture sequence had been completed, and the first views were received later in the day.

* There was a period, following 1963, when Cape Canaveral was renamed Cape Kennedy, but the local inhabitants were not in favour, and eventually the name was changed back again.

Altogether Mariner 4 sent back twenty-one pictures of Mars, some of which were almost blank while others showed considerable detail. Although only one per cent. of the total surface of Mars came under scrutiny, the results provided astronomers with plenty of food for thought, and one of the pictures—the eleventh—was very clear (Plate II); it showed an area in Atlantis, the ochre region between Cimmerium and Sirenum, and revealed a 75-mile crater. By then, of course, the presence of craters had been established, and more than seventy were finally recorded. This is no place to go into details of how the television techniques were applied, or of the 'computer enhancement' in which the raw pictures were electronically dismembered, cleaned up, and reassembled to bring out the details. Suffice to say that the techniques evolved were more or less new, and the power actually received from the space-craft was a tiny fraction of one watt: to be precise, 0·000000000000000001 watt (if you care to count the number of zeros, you will find that there are eighteen of them). Anything of the kind would have seemed hopelessly futuristic even at the start of the Space Age, less than a decade earlier.

There was an obvious temptation to compare the Martian surface with that of the Moon, but even at that early stage it was clear that to take the analogy too far would be unwise. I well remember a comment made by one of the NASA scientists when asked whether Mars was like the Earth or like the Moon. He paused, and then said "Well—to me, it's like Mars," which was a very fair summing-up. But there were some really vital conclusions to be drawn immediately. First, the dark areas and the ochre tracts did not seem to be very different except in colour; there were craters in both. Some were seen in the famous dark region of the Mare Sirenum, and there were others in Phlegra, Phæthontis, Atlantis and elsewhere. Neither were the boundaries of the dark regions as well-marked as had been expected, and it is probably true to say that these first Mariner 4 pictures finally killed off the vegetation theory which had been regarded as so nearly proved. Also, Mars was not a world with a flattish landscape, and one elevation on the famous eleventh frame was estimated to rise to 13,000 feet (though the existence of giant volcanoes was not then suspected). Yet one inference drawn at the time has been found to be wrong. It was supposed that the formations were very ancient and considerably eroded, which now seems to be the very reverse of the truth.

Despite the excellence of the television pictures—by 1965 standards, that is to say—the most important results concerned the atmosphere, which turned out to be very thin indeed. The method adopted was most ingenious, and has always been called the Occultation Experiment. Shortly after 02·19 hours on 15 July, more than two hours after closest approach, Mariner 4 went directly behind Mars. Needless to say, it could not be seen; to glimpse a space-craft from Earth at such a range would mean using a telescope much more powerful than anything we can build. But radio signals were clear, and just before the actual occultation these signals were naturally coming to us after having passed through the Martian atmosphere. The way in which these signals were affected gave reliable clues as to the atmospheric composition and density— and there was another opportunity as Mariner emerged from behind Mars at 03·13 hours. The occultation itself had therefore lasted for $1\frac{1}{4}$ hours, while the signal distortion due to the Martian atmosphere made itself evident for about two minutes before occultation and another two minutes after emersion.

The occultation took place on the sunlit side of Mars, between the regions named on the maps as Electris and Mare Chronium—latitude 55° south, longitude 177° east. The signal was picked up again as Mariner came out on the night side above Acidalia Planitia (Mare Acidalium), at 60 degrees north 44 degrees west, just before sunrise over that part of Mars. The fly-by had been brief indeed, and there could be no second chance.

To be candid, astronomers were rather depressed by the results. It seemed that the ground pressure on Mars could be no more than 4 to 7 millibars, or about 1% of the pressure of the Earth's air at sea level. The main constituent was the heavy gas carbon dioxide, though it was thought possible that there might be a considerable amount of argon as well. This led at once to the revival of the Ranyard-Stoney theory of polar caps made up of solid carbon dioxide rather than ordinary ice, and it was not until the Viking missions of 1976 that the pendulum swung back again.

Mariner 4 had done all, and more, than its makers could have hoped of it. Now, of course, all track of it has been lost, but no doubt it is still in orbit round the Sun, and we can be reasonably confident about its path, particularly as it was re-contacted for a while in 1966. Mariner 4 is completing one circuit of the Sun every 587 days, with a distance ranging between 102,531,000 miles at perihelion out to 146,029,000 miles at aphelion, not very different from the probable orbit of the dead Mariner 3. Undoubtedly it has an honoured place in astronomical history.

I need do no more than mention Russia's Zond 2, another intended fly-by, which was launched two days after Mariner 4 but which was yet another failure. Probably it passed within a thousand miles of Mars around 6 August 1965, but long before then it had ceased to transmit, and it was some years before the Russians made another attempt. Meantime, events had been taking place much nearer home. The Apollo programme had been put under way, and, as almost everyone will remember, the first astronauts landed on the Moon in July 1969. Only a few days after Neil Armstrong's never-to-be-forgotten 'small step' on to the Mare Tranquillitatis, there was more news from Mars, this time from Mariners 6 and 7. (In case you are wondering about Mariner 5, I should explain that it was a successful Venus probe. It by-passed the planet in October 1967, but need not concern us here.)

Mariner 7 was launched more than five weeks after its predecessor, but took only 130 days to reach Mars, as against 156 days for Mariner 6. The distances covered were, respectively, 197,000,000 miles and 241,000,000 miles. The orbits were of the now-conventional transfer pattern; each probe needed only a single mid-course correction, and there was only one serious alarm. This was near the end of the voyages, on 30 July 1969. Mariner 6 was within seven hours of its pass, with Mariner 7 not far behind, and everything seemed to be under control when, with no prior warning, Mariner 7 'went silent'. The Deep Space Tracking Station at one of the key stations, Hartebeespoort in South Africa, reported loss of signal, and it was believed that the probe had been hit by a meteorite large enough to damage it. Moreover, Mariner 7 had lost its Canopus lock, which explained why it had gone out of contact. Fortunately it was still able to receive commands from Earth, and eventually Canopus was re-located by its sensors, so that signals were picked up once more. It was found that there was a certain amount of damage, but nothing serious, and the trajectory had been changed very slightly—enough to make the closest approach to Mars occur ten seconds later than had been predicted. After a journey lasting for over four months, this was not very much! In the end, the television pictures from Mariner 7 proved to be the more detailed of the two.

Each probe took what are called far-encounter pictures while still a long way from Mars (well over half a million miles in some cases), and these showed the south polar cap excellently, though without any conspicuous dark band round its edge. Also shown was a bright ring—the Nix Olympica (Olympus Mons) which had been seen by Schiaparelli and all later observers of the planet, though neither of the 1969 probes could indicate that it is really a 15-mile-high volcano. It was, I suppose, these long-range views which gave the final coup de grâce to Lowell's canals in any form.

Altogether, Mariner 6 took 50 far-encounter and 25 close-approach pictures; Mariner 7, 93 and 33 respectively. The areas covered were not the same, because the first probe passed more or less over the Martian equator while Mariner 7 concentrated upon the extreme south. Among the regions shown in the pictures were the Mare Sirenum, Sinus Sabæus, Sinus Meridiani, Deucalionis Regio and Hellas. Picture resolution was much better than with Mariner 4, partly because of improved techniques and partly because the 1969 space-craft went closer to Mars, passing a mere 2200 miles from the surface.

There is no point in giving further details about the orbits, and it is enough to say that after rendezvous each probe continued in a closed path round the Sun, so adding to the growing swarm of tiny artificial planets. Generally speaking, the results of the missions were much as had been expected. The low atmospheric density was confirmed; Mariner 6 reported that the pressure in the Sinus Meridiani region was 6·5 millibars, while Mariner 7 gave a mere 3·5 millibars over the area which had been known as Hellespontica Depressio, indicating that it was not a depression at all. At noon on the equator the temperature was found to rise to +60 degrees Fahrenheit, falling to below −100 degrees at night. Mariner 7 made temperature measurements of the south cap area, and found a minimum of −190 degrees Fahrenheit, which was very close to the frost point of carbon dioxide there. This agreement was taken as strong circumstantial evidence that the cap really was made up of solid carbon dioxide rather than water ice.

One of the most interesting features to come under scrutiny was Hellas, no longer regarded as a plateau, but as a basin. Virtually no details could be seen inside it, whereas the neighbouring regions—including the Hellespontica 'Depressio' —were thickly cratered. Craters were also seen in the regions covered with white polar deposit, and in particular there was a striking picture of a double crater arrangement which was at once nicknamed the Giant's Footprint.

After the Mariners had completed their passes and gone on their way, a sense of anti-climax prevailed. So far as could be made out from the 20 per cent. of Mars photographed during the close-range encounters, there were three types of terrain: cratered regions, what were known as 'chaotic' areas containing few craters but many jumbled ridges, and the much less common smooth parts of the surface of which Hellas was the best example. Astronomers tended to dismiss Mars as a rather dull, inert world with no features of striking interest, and it was even said that the surface was more like that of the Moon than had been expected. Incidentally, no trace of the famous Violet Layer was recorded by the instruments on board the probes, and neither was there a detectable magnetic field, while radiation belts around the planet were conspicuous only by their absence.

There was another factor, too. Because of the prevailing view that the caps were made up almost entirely of solid carbon dioxide, and the certainty that much of the atmosphere was composed of carbon dioxide gas, the possibility of any kind of Martian life receded into the background, particularly since there was no longer the slightest chance that the dark regions were vegetation-tracts. I have always thought it rather fortunate that plans for two more Mars probes, Mariners 8 and 9, were already so well advanced that there was no thought of cancelling them.

We now know that by pure ill-chance, the 1969 space-ships passed over some of the least spectacular areas on the whole of

Mars. They missed the volcanoes, the riverbeds and the gaping chasms, so that they gave an entirely misleading impression of what Mars is really like. This was not the fault of the planners, but it does show the folly of jumping to conclusions. Within less than three years, all our ideas were destined to change.

Chapter Nine

MARINER 9—AND OTHERS

THE FIRST PHASE of direct Martian exploration ended with the probes of 1969. The second phase began with the space-ships of 1971, of which there were four: two American and two Russian. Their aims were not identical. The United States craft were designed to go into closed orbits around Mars and to map the surface as thoroughly as they could, naturally obtaining much miscellaneous information at the same time. The Soviet planners were more ambitious; they hoped to carry out soft landings, and to pick up transmissions direct from the planet.

In the event, only one of the four vehicles was successful. This was Mariner 9, which more than made up for the failures or near-failures of the other three. Before the Viking programme, indeed, Mariner 9 was the source of practically all our reliable information about Mars, and I must discuss it in some detail, so it may be as well to clear the way by disposing of the other space-craft first. This is logical enough, because Mariner 9 was actually the last to be launched even though it led the race to arrive.

Mariner 8, sent up from Cape Canaveral on 8 May 1971, did not arrive at all. The second stage of the Atlas-Centaur launcher failed to ignite; the probe made an undignified descent into the sea some 350 miles north-west of Puerto Rico, and that was that.

Next it was Russia's turn, and the Soviet rocketeers sent up Mars 2 on 19 May, following it with Mars 3 on the 28th. As usual, the launching procedures were carried through without a hitch. Both the probes were put into correct orbits, and both duly rendezvoused with Mars more than six months later. They were, incidentally, the heaviest vehicles so far sent there. Each weighed 10,250 pounds, as against 2150 for Mariner 9.

The Russian vehicles were made up of two parts each: an orbiter and a lander. Basically, the idea was to put the combined craft into orbit round Mars and then, at a prearranged moment, separate the landing section, which would descend through the Martian atmosphere and slow itself down partly by rocket braking and partly by parachute, finally coming to rest gently enough to avoid damage. Meanwhile, the orbiter would continue in its stable path, acting as a relay for the lander as well as carrying out investigations on its own account. I will say much more about this scheme when we come to Viking, so there is no need to dwell upon it now, particularly as neither of the Russian vehicles worked properly.

Mars 2 entered its closed orbit successfully on 27 November 1971, after a total journey of 292,000,000 miles. Just before doing so it ejected a capsule which apparently hit Mars about 300 miles south-west of Hellas, in latitude —44·2 degrees, longitude 313·2 degrees. The capsule carried a Soviet pennant. Whether it carried anything else was not stated, and in any case no scientific information was obtained from the landing section of the probe, so that the experiment would seem to have been fairly useless. All the same, the pennant does represent the first man-made object to land on Mars (just as the Russian Luna 2 was the first object to hit the Moon, in September 1959; how long ago that seems!).

On 2 December, Mars 3 followed its predecessor into orbit, and it too ejected a capsule, which came gently down and landed at latitude —45 degrees, longitude 158 degrees, in a rather light-coloured region between Electris and Phæthontis not far from the northern limit of the south polar cap. Within a minute and a half it began to transmit a picture from the Martian surface. Momentarily all seemed well, but then, after a mere twenty seconds, transmission broke off. It was never re-established, and the fragment of picture received in the U.S.S.R. showed no detail whatsoever. The 12-foot lander had entered the Martian atmosphere at 13,500 m.p.h., and there is no reason to doubt that the actual touch-down was accomplished without mishap, but something unpleasant happened immediately afterwards. Remember, a violent dust-storm was raging at the time, and this may have been the cause of the trouble. Soviet scientists suggested that the parachute, which had been jettisoned at 100 feet above the ground, landed on top of the probe. Alternatively, the lander may have been toppled over, or sunk into windblown dust. All these theories were put forward, though in view of the later Viking revelations that Mars is rockstrewn I suggest that the luckless lander came down on top of a large boulder and was fatally holed. No doubt we will find out in the future, when the first expedition goes to the area and locates the remains. (I refuse to accept yet another theory—that a little green Martian wandered over to the probe and switched it off!)

To be fair, the two Russian vehicles were not total failures, because their orbiting sections did send back some information. They carried radio telescopes operating at a wavelength of 3·4 centimetres, and found that the temperature half a metre below the ground never rises above a temperature of —40 degrees Centigrade; they also reported that some of the clouds rose to heights of over six miles, that there were mountains rising to 10,000 feet and depressions sinking to 4,000 feet below the mean surface level, that the ochre tracts were likely to be covered with sand rather than limonite or some such mineral, that there was no detectable magnetic field, and that the previous estimations of the atmospheric density were correct. Values of between 5·5 and 6 millibars were given, so that the Martian atmosphere seemed to have about 1/200 the density of our own. The estimations of surface heights were drawn from measurements of the amount of atmospheric carbon dioxide lying above the various regions. Most of the results were of the right order, though of course we now know that some of the peaks are vastly higher than 10,000 feet.

By this time Mariner 9, dispatched from the Cape on 30 May,

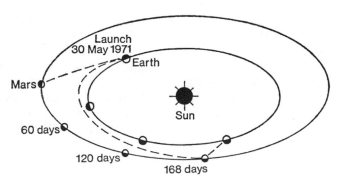

Fig. 29. Flight path of Mariner 9

was already in orbit (Fig. 29), and had started to send back information, though at first it could do no more than show the tops of dust-clouds. (It had also taken pictures of the two Martian satellites, Phobes and Deimos, but I would prefer to reserve these for a later chapter.) Mariner, of course, had no landing section, so that it was 'all orbiter'. On 14 November its rocket motor was fired for 15 minutes 23 seconds, slowing the space-craft down and putting it into its circum-Martian path. The initial revolution period was 12 hours 34 minutes, slightly more than half a sol, but it was then changed to 11 hours 58 minutes 14 seconds, which brought the probe down to a minimum distance of 850 miles from the surface. The very first pictures showed four spots which later proved to be the tops of volcanoes poking through the dust. We recognize them today as Olympus Mons, Ascræus Mons, Pavonis Mons and Arsia Mons, the giants of the Tharsis ridge. They had been known earlier under the names of Nix Olympica, Ascræus Lacus, Pavonis Lacus and Nodus Gordii.

When the dust cleared, toward the end of the year, Mariner 9 could begin its main work. On 30 December the orbit was altered again, giving a new period of 11 hours 59 minutes 28 seconds. Transmissions continued until 27 October 1972, and altogether Mariner sent back 7,329 pictures, covering practically all of the planet, as against the puny 10 per cent. recorded from the two space-ships of 1969. I doubt whether anyone seriously expected the pictures to be as good as they actually were; they were breathtaking, and without them the even more ambitious Viking missions could never have 'got off the ground', either metaphorically or literally.

Mariner 9 is still circling Mars, though it is now dead. It will stay there for some time; it is suggested that the drag caused by the thin Martian atmosphere will eventually cause it to crash-land, but certainly not for fifteen years at least, and probably much longer. It is interesting to note that at the time when these words are being written (1977) Mars has a grand total of nine moons, of which only two are natural.

Before indulging in speculation, let us summarize what Mariner 9 told us about the surface topography of Mars.

When it had entered its 'mapping path', which carried it from 1025 miles out to 10,610 miles from the Martian surface during each revolution, Mariner 9 had to wait patiently for the dust to clear. As we have noted, the first features to be shown were the volcano-tops, Nix Olympica (Olympus Mons) and what were first called North, Middle and South Spots (Ascræus, Pavonis and Arsia). It is worth recalling that on the occasion of another dust-storm, many years earlier, Schiaparelli had found that his 'Nodus Gordii' and 'Olympic Snow' were almost the only features to be seen. He guessed, correctly, that they must be high, so that we may forgive him for classifying Pavonis and Ascræus as lakes!

Once the Martian atmosphere had cleared, and full-scale photographic coverage could begin, it was obvious that the two hemispheres of the planet are not alike. It may be said that they are divided by a great circle inclined at about 50 degrees to the equator, but for convenience I propose to term them 'north' and 'south'. The southern hemisphere is heavily cratered over much of its extent, though it does contain the almost featureless basins of Hellas and Argyre. There are abundant craters in the dark regions of Sinus Sabæus and Tyrrhena Patera (Mare Tyrrhenum) as well as in other places. The northern hemisphere contains fewer of these thickly cratered areas, though in some regions they do exist; there are more volcanoes, and almost the whole of the great Tharsis volcanic chain lies in the northern hemisphere, with only Arsia Mons slightly south of the equator. Measurements from Mariner 9 showed that in general the southern hemisphere lies about $1\frac{3}{4}$ miles above the nominal mean radius of the planet, with the northern hemisphere lower as well as smoother. Mars is thus slightly pear-shaped, which has important effects upon the changing tilt of the planet's axis.

No doubt the difference between the two hemispheres of Mars is highly significant, and I will have more to say about it later. As a preliminary comment: there is absolutely no doubt that the volcanoes in Tharsis and elsewhere really *are* volcanoes of the 'shield' type, very similar to our own Hawaiian volcanoes though on a larger scale (Fig. 30). It has been suggested, very reasonably, that they are relatively young, and that their lava outpourings have masked many of the older craters which used to exist nearby. It has also been maintained that most of the craters themselves, apart from those on the volcano crests, are

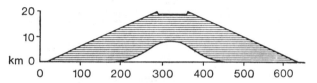

Fig. 30. Relative sizes of Olympus Mons (shaded) and the
most massive Hawaiian volcano (white)

of impact origin—that is to say, formed by the descent of meteorites on to the Martian surface. According to this idea, even
Hellas and Argyre are impact basins. I am frankly sceptical.
Impact craters must exist on Mars, but I hold the view that
most of the large structures are of internal origin. This is a
personal opinion, but let us leave the matter there for the
moment and take a closer look at the Martian scene, beginning
with the north pole and then considering each quadrant. In the
maps given here I have put north at the top, because Earth-
based observers will see none of the details shown apart from
the dark albedo regions and the main basins, plus a few of the
volcanic structures such as Olympus Mons.

North Polar Area, down to around latitude +60 degrees. This is
covered with white deposit during winter. There are not very
many craters by Martian standards, but some are visible, and
one of them (Korolev) was later shown by Viking to be filled
with ice. The main feature of the area is the layered or lami-
nated terrain. This is made up of what may be called 'staircases',
very thin layers of alternate light and dark hue, whose gently
sloping faces show a certain amount of relief, descending from
the pole toward the cratered regions. The layers have been
likened to saucers of various sizes, and are peculiar to the two
polar zones. The huge, dusky Vastitas Borealis extends right
round the planet, from approximately latitude +55 degrees to
+67 degrees, and I suspect that its presence has been partly, at
least, responsible for the reported 'Lowell band' seen at the
time when the polar cap is shrinking.

North-East Quadrant. (Fig. 31) This contains the great Syrtis
Major, now known to be a plateau sloping off to either side. To
the west is one of the few heavily-cratered regions of the

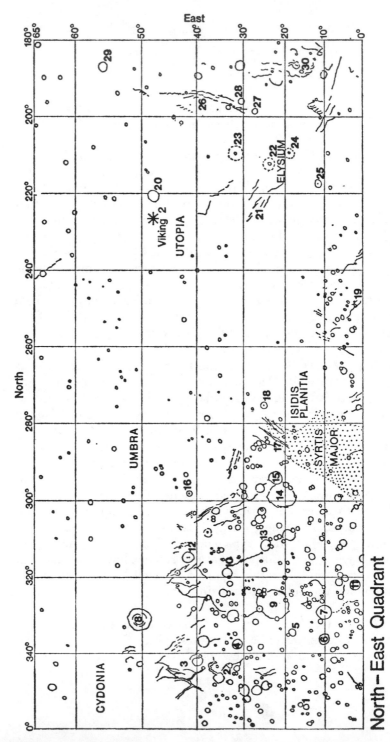

Fig. 31. Main features of Mars; north-east quadrant

North-East Quadrant

123

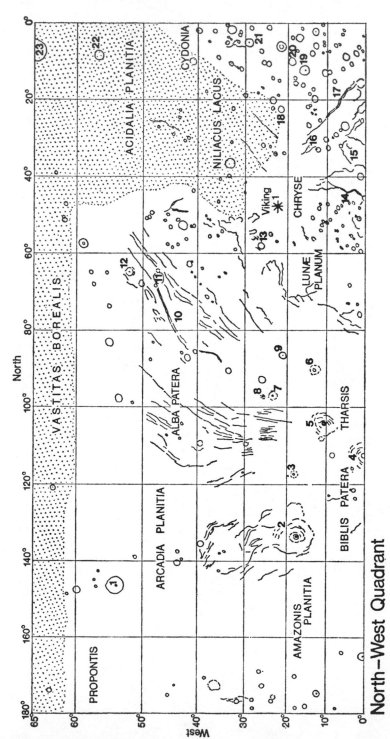

Fig. 32. Main features of Mars; north-west quadrant

North-West Quadrant

1 Milankovitch
2 Olympus Mons
3 Jovis Tholus
4 Pavonis Mons
5 Ascræus Mons
6 Tharsis Tholus
7 Ceraunius Tholus
8 Uranius Tholus

9 Fessenkov
10 Mareotis Fossa
11 Barabashov
12 Perepelkin
13 Sharonov
14 Shalbatana Vallis
15 Simud Vallis
16 Tiu Vallis
17 Ares Vallis
18 McLaughlin
19 Trouvelot
20 Becquerel
21 Curie
22 Kunowsky
23 Lomonosov

125

northern hemisphere, making up ochre tracts such as Aeria, Arabia and Eden. The crater Schiaparelli has actually been seen from Earth as a tiny speck, though naturally nobody in pre-Mariner days had any idea of its true nature (unless we include the extraordinary observations made by Barnard and Mellish). Sinus Meridiani or the Meridian Bay, to the lower left, seems to be an albedo feature and nothing more. The zero for longitude is now taken to be the craterlet called Airy-o in honour of Sir George Airy, the rather formidable last-century Astronomer Royal who was largely responsible for the acceptance of Greenwich as the zero for terrestrial longitudes.

Isidis Planitia (formerly Isidis Regio), bordering Syrtis Major, slopes down to the volcanic region of Elysium Planitia, where there are several well-marked volcanoes: Elysium Mons, Hecates Tholus and Albor Mons, all of which are decidedly lofty even though they cannot compare with the giants of Tharsis. The relative sparseness of crater distribution around here is very evident. The quadrant also includes Utopia Planitia; it was here that the second Viking lander came down, not very far from the prominent crater now called Mie. Utopia had been expected to be relatively smooth. Instead it has proved to be what has been termed 'a forest of rocks'.

I have already commented upon the fact that some of the most famous Martian canals were in this quadrant—Phison, Euphrates, Protonilus, Hiddekel and others. All, alas, were quite illusory.

North-West Quadrant. (Fig. 32) This is the quadrant which contains most of the Tharsis volcanoes. Pride of place must go to Olympus Mons, which has a base 375 miles across and a maximum altitude estimated at 78,000 feet—about 15 miles, which is over twice the height of our Everest above sea level. Olympus is a shield volcano, and is crowned by a 40-mile caldera. It is the largest volcano known to us, and dwarfs anything we find on Earth. During Martian mornings it is wreathed in clouds formed of water ice, condensed from the atmosphere as it cools while moving up the slopes of the volcano.

Shield volcanoes were originally so called because of a supposed resemblance in shape to the shields of early Viking warriors. On Earth they are confined mainly (not entirely) to three areas: Hawaii, the Galapagos Islands and Iceland, plus a few in California and New Zealand. Summit calderas are the rule, and this also applies to the Tharsis volcanoes. Ascræus Mons has a 31-mile caldera; Pavonis, 28 miles; and Arsia Mons, which actually lies just in the southern hemisphere, beats them all with a caldera over 85 miles in diameter.

The Tharsis volcanoes are associated with what seem to be drainage systems. The most conspicuous features of the surrounding area are the canyons, mainly the tremendous Vallis Marineris and the labyrinth at Noctis, which lie in the southwest quadrant. Another interesting feature is the Tractus Albus, which is ridge-like and which runs from Arsia Mons as far as Acidalia Planitia. It is certainly associated with the diagonal line of the main volcanoes, of which the four giants are not the only members.

The Tractus Albus runs between the Tharsis volcanoes to the east and the darkish Lunæ Patera to the west; it drops in height with increasing distance northward. Acidalia is a low-lying region, and is quite unlike the Syrtis Major even though these are the two most conspicuous dark areas on the whole of Mars.

Tharsis slopes off sharply to the west (Amazonis Planitia), but less sharply toward the east. Here we come to an ochre tract, Chryse Planitia, which merges into the very similar Xanthe. It was in Chryse that the lander of Viking 1 made its epic descent.

South-West Quadrant. (Fig. 33) Two major features dominate this part of Mars: the Vallis Marineris, and the basin of Argyre I.

The Vallis Marineris, or Mariner Valley, is huge even by Martian standards. It is 2500 miles long, over 45 miles wide at its broadest point, and probably 20,000 feet deep, which is practically three times the depth of the Grand Canyon of the Colorado. It seems to be in the nature of a rift, and comparisons have been made between it and our Red Sea, though the Vallis is considerably the longer of the two. The tributaries running from it are particularly significant, and it is hard to resist the conclusion that they were formed by the action of running water. The whole system extends from Tharsis right

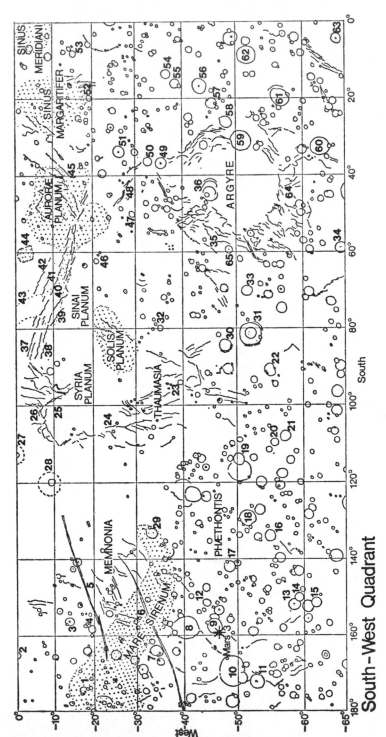

Fig. 33. Main features of Mars; south-west quadrant

through to Auroræ Planum, and one has to admit that it does correspond more or less to the site of the Coprates canal shown on the old Lowell-type maps, though on the whole I fear that this may be pure coincidence. To the north-west is another familiar feature, long known as Juventæ Fons or the Fountain of Youth, which may be an irregular depression and which is visible from Earth as a dark patch. Lowell, needless to say, classed it as an oasis.

Adjoining the Tharsis region and the high area of Phœnicis is the Noctis Labyrinthus. This is a huge system of canyons, unlike anything else so far discovered, and making up the pattern which has been nicknamed the Chandelier. The canyons are not like the rills of the Moon, and may well be due to fracturing of the surface, together with the withdrawal of magma from below. Each of the main canyons has an average width of about a dozen miles. No doubt this region will be one of the main Martian tourist attractions in the centuries to come . . .

To the east are two famous dark areas, Mare Cimmerium and Mare Sirenum; Cimmerium is rather high, while Sirenum seems to have no special characteristic apart from its colour. There are many craters, and according to Russian reports the lander of their probe Mars 3 came down in a position which corresponds to one of these craters, Ptolemæus.* It is a great pity that no useful transmissions were received, because this thickly-cratered area is quite different from the ochre tracts of Chryse or Utopia.

Argyre Planitia (formerly Argyre I, to distinguish it from a rather smaller feature, Argyre II) is a basin of the Hellas type, though it is neither so large nor so deep. Little can be seen in it, but to its north there are some remarkable sinuous features which we must, I feel, class as dry riverbeds. To the north-west lie the dark, cratered regions of Mare Erythræum and Margaritifer Sinus.

* One of the most famous walled plains on the Moon is also called Ptolemæus. There is considerable overlap in the naming of lunar and Martian craters, which will inevitably lead to confusion in the future. The Martian nomenclature has not yet been fully completed, which is why I have had to retain some of the older names such as Mare Erythræum; these will certainly be revised shortly, and the term 'Mare' is bound to be dropped.

South-East Quadrant. (Fig. 34) Overall, this is one of the most heavily cratered parts of Mars, and there are few true volcanoes, though a couple have been located in Tyrrhena Patera, north-east of Hellas at around latitude −22 degrees, longitude 253 degrees. Magnificent systems of river beds are seen in the Rasena region (latitude −25 degrees, longitude 190 degrees). Among the craters, one of the most extraordinary is Proctor, east of Hellas, which appears to be filled with sand-dunes. It lies on the Hellespontus, a cratered region sloping down toward the Hellas basin itself.

I have already said a good deal about Hellas. It is over 1300 miles across, and apart from one small patch, seen by Antoniadi and named by him Zea Lacus, it is virtually featureless. Across it Schiaparelli, in 1877 and 1879, drew two canals which were later called the Alpheus and the Peneus, making up a cross. No trace of them was shown by Mariner 9, so yet again we seem to be dealing with alleged canals which do not exist in any form whatsoever.

South Polar Region. Here we are back to layered terrain, but the region round the pole itself is more thickly cratered than its northern counterpart. Around latitude −60 degrees there is a belt of darkish terrain, and this may explain the 'Lowell Band', though it is less well-marked than the northern Vastitas Borealis. There are also some lighter areas which may be basins: Argyre II, Thyle I and Thyle II, all of which are covered with cap material during the depths of winter.

This account of the scene from Mariner 9 is not meant to be at all complete. All I have tried to do is to give a brief survey of the various types of features discovered. Comparing these results with those of the earlier probes leaves one in no doubt as to the magnitude of Mariner 9's achievement, which was all the more remarkable because of the need to shoulder the programme originally assigned to Mariner 8. From being regarded as an inert, lunar-type planet, Mars was transformed into an active world, with features which defied explanation but which were of surpassing interest. It was also thought possible that instead of being in the last stages of senility, Mars was gradually heating up and preparing to enter a new and much less hostile phase in its evolution. There were also various

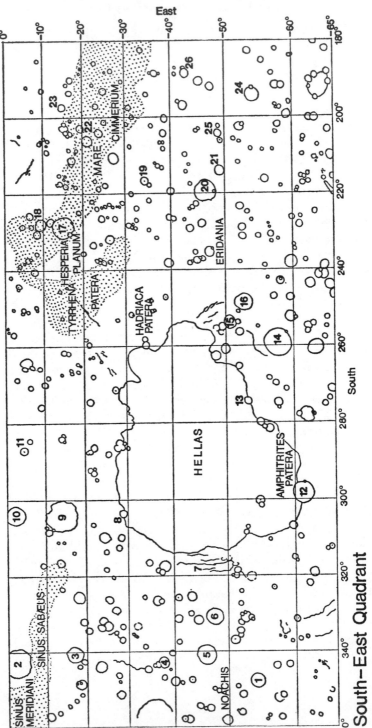

Fig. 34. Main features of Mars; south-east quadrant

133

enigmas. The river beds did not look ancient, and they were not seriously eroded; yet there can be no water on Mars today in liquid form. It was all very puzzling.

There were many regrets when Mariner 9 finally came to the end of its career, but already the plans for Viking were being drawn up, and in the meantime there was plenty of analysis to be done. Also, the Russians had certainly not lost interest, and in the summer of 1973, less than a year after the last signals from Mariner 9, they dispatched a positive fleet of space-ships. Four vehicles were sent up in rapid succession. At least two were of the orbiter-plus-lander type, and much was expected of them, particularly as the Soviet rocket planners had already had such marked successes with Venus—which is a far more difficult subject for exploration than Mars.

Sad to say, the Soviet fleet was a failure, and added very little to the information drawn from Mariner 9. To avoid tedious repetition, it may be best to treat the four probes briefly and in order of launch:

Mars 4. Launched 21 July 1973. On 10 February 1974 it approached its target, but its braking engine failed to operate, and the probe missed Mars by over 1300 miles, so that it continued on its way in a solar orbit. A few television pictures were obtained during the involuntary fly-by, but their quality was poor.

Mars 5. Launched 25 July; approached Mars on the following 12 February. This time the braking engine worked, and some television pictures were obtained. They showed craters and river beds, but with nothing like the clarity of Mariner 9.

Mars 6. Launched 5 August; reached the neighbourhood of Mars on 12 March 1974. The lander was successfully broken free from the orbiter at a distance of almost 30,000 miles from Mars, and the initial descent seemed promising. Rocket braking was used for five minutes, and then the main parachute was deployed. The parachute phase lasted for 148 seconds, but then contact was permanently lost. So far as is known, the lander came down at latitude −24 degrees, longitude 25 degrees west, between Mare Erythræum and Margaritifer

Sinus, but nothing more was heard from it. Some television pictures were received before the separation, and the Soviet news agency said that from Mars 6 the planet looked like a red, waning Moon.

Mars 7. Launched 9 August 1973; approached Mars on 9 March 1974, three days before Mars 6. This time the lander separated from the orbiter prematurely, and the vehicle missed Mars by 800 miles, moving uselessly into an orbit which will continue to take it round the Sun.

The story of the Mars Fleet is rather depressing, and since then the Russians have left the Red Planet alone. But in Cape Canaveral, the Vikings were being made ready.

Chapter Ten

THE VIKINGS

IT HAS BEEN SAID that the Viking missions were the most ambitious ever undertaken by the American space-planners up to 1977. True, a little of the novelty had been stolen by the Russian success in landing two capsules on the surface of the much more hostile planet Venus, and managing to receive one direct picture from each; but the Martian Vikings were expected to go on transmitting for many weeks rather than for an hour or so. Moreover, there was the all-absorbing problem of the search for life.

What Mariner 9 had not done, and could not possibly do, was to tell us whether or not Mars is completely sterile. This was one of the main tasks for Viking, but let me stress at the outset that it was not the sole reason for attempting a controlled landing. Indeed, many scientists maintained that some of the other investigations, such as analysis of the soil materials and atmospheric composition, were even more important. Opinions differed, but everyone agreed that whatever happened the Vikings would—if successful—usher in a new phase in our exploration of the Solar System.

Two missions were planned. Vikings 1 and 2 were identical, and were intended to carry out identical tasks from two different regions of the planet. No. 1 was aimed at the 'Golden Plain', Chryse, more or less between the Margaritifer Sinus to the south and the dark mass of Acidalia to the north. No. 2 had as its target Cydonia, an area considerably closer to the north pole, and just about at the limit of the northern polar cap at its maximum spread. The sites had been chosen very carefully. They were relatively low-lying, and were expected to be reasonably smooth; also, they were well placed to receive any residual moisture, because they lay at the end of the great drainage-system associated with the Tharsis volcanoes. If there were any moisture on Mars, it would be expected to be there.

Each vehicle consisted of two main parts, an Orbiter and a Lander (Figs. 35 & 36). Naturally, the two would travel together across interplanetary space, and would be put into a closed path around Mars in much the same way as for Mariner 9. At the planned moment the Lander would be separated, and would descend through the Martian atmosphere to make a gentle descent on to the surface, leaving the Orbiter to its own devices. Not that the Orbiter was at the end of its career: far from it. Not only would it carry on with its own investigations, but it would also act as a relay to send back messages from the Lander. Indeed, without the Orbiter, the Lander would have been very restricted in its ability to communicate with Earth.

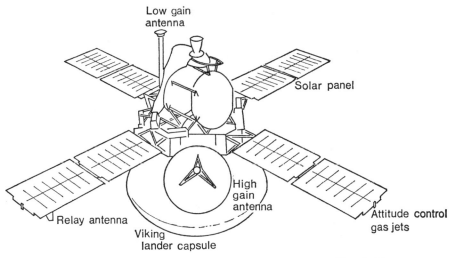

Fig. 35. The Viking vehicle. (Vikings 1 and 2 were identical.)

The launching vehicle—a Titan 3/Centaur rocket combination—had been well tested, and was expected to give no trouble. Neither was there any reason to doubt that there would be any real difficulty in putting the Viking into its path round Mars. The main risk was in the landing itself, which had to be completely automatic. Once the order to separate had been sent from Earth, the landing manœuvre could be neither stopped nor modified. Moreover, a time-lag would be inevitable, because radio waves would take over nineteen minutes to reach the Earth. When Viking 1 touched down, the distance between Mars and ourselves was 212,000,000 miles, while at the time when Viking 2 made its descent Mars was even further away.

In view of what they accomplished, these first two Vikings were amazingly small. The Orbiter was octagonal, 8 feet across, 10·8 feet high and a mere 32 feet across the full spread of its solar panels. The Lander was 10 feet across and 7 feet high, with a total unfuelled weight of 1270 pounds. It was a strange-looking, three-legged device (Fig. 36), and everything

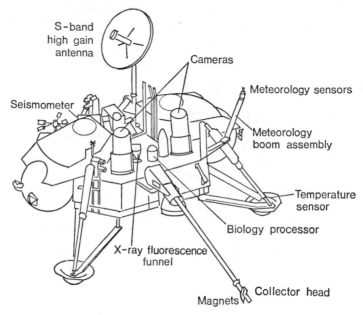

Fig. 36. Viking Lander. (The biology box, gas chromatograph—mass spectrometer, X-ray fluorescence spectrometer and pressure sensor are internally mounted.)

depended upon its coming down gently and at an acceptable angle. A tilt of more than 19 degrees would put it out of action insofar as transmissions were concerned, and if the vehicle landed upon a large boulder it would be fatally damaged; the clearance of its base from the ground was a mere 8·7 inches. Unfortunately, boulders of 'dangerous' size were below the resolving limit of either Orbiter photography or Earth-based radar measurements, so that a good deal of luck was needed. Before the actual landings, it is fair to say that the scientists at the Jet Propulsion Laboratory at Pasadena (California) estimated the chances of success at no more than 50-50.*

It has been claimed that the Lander is the cleverest machine ever built by Man. This may be true, but it was also extremely delicate, and it had to be protected as well as possible. It also had to be sterilized, because to carry any Earth contamination to Mars would be scientifically disastrous. One way to kill any organism is by intense heat, and before being sent up the Lander was accordingly 'cooked' for forty hours in an oven heated to well above the boiling-point of water. There is every reason to hope that the treatment was effective. Of course, one can never be absolutely sure, but the risks have been eliminated as far as humanly possible, and no doubt this also applies to the Russian descent capsules.

Politics can never be kept out of science, unfortunately, and it is a fact that several of the most memorable space-shots have been timed for reasons that are not, strictly speaking, scientific. Thus the Space Age began on 4 October 1957, with a Russian satellite: the fortieth anniversary of the Bolshevik revolution. Luna 3, the first probe to go right round the Moon and send back pictures of the far side, was dispatched on 4 October 1959. With Viking, it was hoped to make the initial landing on 4 July 1976, American Independence Day, but at an early stage it became clear that this could not be done. There was trouble before the launching, and Viking 1 was delayed. Eventually the two vehicles were reversed, so that the probe which should have been Viking 2 became Viking 1 (and will so be called hereafter). To the great credit of all concerned, no attempt was made to speed things up. This might have jeopardized the whole mission, and the delay was accepted, though doubtless with a certain amount of regret. Actually, a further delay was encountered later on, because the original landing site had to be rejected, but by then the Independence Day target had been missed in any case.

* A day or so before Viking 1 landed, I presented a television programme about it, together with Dr. Garry Hunt and Professor Geoffrey Eglinton. We had models of the Lander, and we demonstrated what would happen with touch-downs on various types of Martian surface. As a member of the studio team commented, we did at least prove one thing. "If you drop them from a great height," he said dryly, "they break!"

Viking 1 finally took off on 20 August 1975, and Viking 2 followed on 9 September. I do not propose to say much about the launching itself, or about the journey through space; it is enough to note that there were no major hitches. On 19 June 1976 Viking 1 entered a closed orbit round Mars, and work began immediately. (By then both the Martian satellites had been photographed, but I propose to leave all discussion of these odd little worldlets until Chapter Thirteen.)

Unlike Mariner 9, the Viking did not immediately set out to study the whole surface of Mars. Its first task was to check on the Chryse landing site in order to make sure—so far as possible!—that it really was a suitable place. But even before entering orbit, it had sent back some truly magnificent far-encounter pictures which surpassed anything that Mariner 9 had been able to do. The ochre tracts, the dark areas, the basins and the volcanoes stood out splendidly, and the improvements in photographic technique were obvious at once.

The pictures which came back soon after Viking had begun its main programme were absolutely staggering. One view, of the towering Olympus Mons, was obtained on 31 July from a distance of 5000 miles; it was mid-morning on Mars, and the volcano was wreathed in clouds which extended up the flanks to an altitude of at least twelve miles. There could be little doubt that these clouds were of water ice, condensed out of the atmosphere as it cooled while moving up the slopes of the volcano. The giant caldera, formed by the collapse of the volcano top, was splendidly shown above the uppermost clouds. Another picture, taken later, showed the even larger caldera of Arsia Mons, one of the other chief members of the Tharsis group. Then there were smaller features such as the well-formed, 11-mile crater now called Yuty, in the Chryse area not too far from the intended landing-site. Yuty is remarkably lunar in aspect, with regular, terraced walls and a massive central peak crowned by a pit (Plates VII and XI).

On 3 July Viking sent back a picture of part of the tremendous Vallis Marineris. From a range of 1240 miles, the Valley was seen to have destroyed part of the wall of a well-formed crater, and the effect was unlike anything seen before. Remember, the Valley is well over a mile deep, and dwarfs all terrestrial canyons. (On this picture there is a strange-looking black ring (Plate V). This ring is shown on many of the Viking I pictures, but I hasten to add that it is due to a slight defect in the optical system, and is in no way Martian.) Then there were the fault zones, causing valleys which are widened partly by collapse and partly by mass wasting, i.e. the downslope movements of rocks due to gravity. Also of special interest were the subsidence features which could well have been produced by the melting of ice below the surface; there were sand-dunes, such as the dune field near the north wall of the Ganges Chasma; there was a canyon leading into the Vallis Marineris complex; there were extensive lava-flows, and there were one or two amazing pictures, one of which showed a mountain clump which, by chance lighting and shape, gave an uncanny resemblance to a human face! On 11 July Viking obtained a good oblique view of Argyre Planitia, the large Hellas-type basin surrounded by thickly cratered terrain. One feature of this photograph was the presence of detached layers of haze, 15 to 25 miles high (Plate XI).

River beds were much in evidence, and by now there were very few scientists who believed that they could be anything other than dry watercourses. All in all, the Viking pictures added even more interest to what Mariner 9 had already told us, but the main objective was always to study Chryse, where the Lander was due to touch down.

Alarm signals had been sounded from Arecibo in Puerto Rico, where the immense radio telescope, built in a natural hollow in the ground, had been engaged in radar mapping of the Chryse area. There seemed to be a greater degree of roughness than had been expected, and the Orbiter pictures added to the general unease, because quite apart from the desirability of coming down at an angle of less than 19 degrees it was also essential to avoid landing upon a rock which would puncture the space-craft and put it out of action. Eventually the original site was abandoned, and a second target area selected, still in Chryse but rather to the north-west. This seemed to be better, but the final choice was still further west, and the die was cast. On 20 July, the great attempt began.

Absolute precision was impossible, and all that could be done was to aim the probe as accurately as possible and hope for the best. There was every prospect that it would land somewhere inside a restricted 'ellipse of uncertainty', and there were no detectable boulders within range, though the planners were unpleasantly conscious of the fact that neither the radar nor the Orbiter mapping could show features which would still be large enough to cause disaster.

The capsule that separated from the Orbiter was made up of three main sections: the Lander itself, a cone-shaped aero-shell made up of aluminium alloy, and a base cover. The

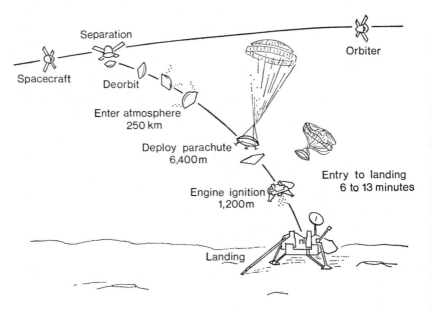

Fig. 37. Descent sequence of the Viking Lander

Lander was enclosed, while the braking engines were part of the aeroshell and the parachute was in the cover. The whole procedure was different from that of a Moon landing, because the atmosphere of Mars, thin though it may be, is dense enough to cause definite friction, which involves heat. It also means that parachutes can be used, though they cannot cope with the full slowing-down process on their own (Fig. 37).

When the descent capsule left the Orbiter, it began its gradual 'coast down' to Mars. It continued to do so, unhampered by any atmospheric drag, for more than three hours, but at an altitude of 800,000 feet above the surface—that is to say, around 150 miles—the atmosphere started to make its presence felt. In preparation for this, the capsule had been turned so that the aeroshell and its heat shield faced the direction of travel. By now the capsule was moving at 10,000 m.p.h., which is not far short of three miles per second, and as it dropped lower and lower the frictional heating built up. The heat shield proved adequate: the deceleration reached its maximum value at 15 to 18 miles above ground level, and for a brief period the path of the capsule levelled off into horizontal flight because of the aerodynamic lift provided by the capsule. With continued slowing-down, the heating became less violent, and the descent could be resumed. At 19,000 feet from the Martian surface the velocity was a mere 1000 m.p.h., and by now, of course, the capsule was well below the level of the tops of the Tharsis volcanoes. This was the moment for the parachute to be deployed, and seven seconds later the aeroshell, its work done, separated from the Lander and drifted away, to fall to the ground well away from the main site.

The last part of the descent was, obviously, the most critical of all, and the scientists back on Earth could know nothing about it, because of the 19¾-minute delay in the reception of signals. In fact the parachute worked perfectly, and within a minute the Lander's velocity had dropped so much that the effects of the winds on its horizontal travel were measurable. The legs were extended to the landing position, and when the radar altimeter gave the height at a mere 3900 feet the parachute was jettisoned. At the same moment the three terminal descent engines of the Lander itself came into play, and seconds later the 440,000,000-mile journey was over. The final touchdown speed was less than 6 miles per hour. Only an hour and a half later, scientists in the control centre at Pasadena were examining the first pictures ever to be successfully transmitted from the surface of Mars.

The touch-down position was a mere twenty miles from the planned impact point, well within the 'ellipse of uncertainty'. And yet luck had played its part. The Lander came down only 25 feet from a boulder which was ten feet across and three feet high—more than big enough to have caused fatal damage if the landing had been made on top of it. As soon as I saw the picture of it, I was reminded of the comment made in the television studio when we were demonstrating with models: "If you drop them from a great height, they break!"

The very first picture, taken immediately after touchdown, showed that the entire landscape was strewn with rocks. One, near the centre of the picture, was four feet across, and six feet away from the Lander. One pad of the third leg was shown, and had penetrated only 1·4 inches into the soil, so that clearly the Martian surface was reassuringly solid, even though it was later found that another of the Lander's legs was covered with sand. The overall impression was of a barren, rocky desert, with extensive sand-dunes as well as pebbles and boulders. (Plate XII, *top*).

What, then, about the colour? On the sol after the landing, Viking sent back a colour picture showing that the fine, granular material so much in evidence was rusty red, while most of the rocks were of similar hue, so that presumably the red material—probably limonite (hydrated ferric oxide)— formed a thin veneer over the darker bedrock. One major surprise was the colour of the sky. The first pictures showed it to be bluish, but this proved to be wrong; corrections were made, and the sky showed up as salmon-pink. Apparently the pinkness is caused by the innumerable fine dust-particles suspended in the atmosphere, and which absorb the blue light from the Sun. The pioneer astronauts will find the scene unearthly in every sense of the term.

As the days (and sols) went by, more and more pictures came through, and each provided new topics for discussion. The colour pictures were naturally the most spectacular, though it must be stressed that these are actually computer reconstructions of three black-and-white images taken through filters. The main discussions centred round the rocks, which were seen to be of various types. Some were coarse-grained and knobbly, while others were lighter-coloured and irregularly shaped, and there were many blocks of considerable size. There was little doubt that the rocks were of volcanic origin. Superficial comparisons are always dangerous, but recently I looked first at a black and white picture from Chryse and then at a black and white photograph I had myself taken in Iceland some years ago. It was not easy to tell which was which. I also looked at a colour photograph of the Tibesti Desert, about a thousand miles south-west of Cairo; here the resemblance to Chryse was even more striking—redness and all.

All round the Lander there was abundant evidence of wind-blown material, and some of the rocks were obviously smoothed, but others were sharp, and this caused a certain feeling of surprise. Meanwhile, a great deal of extra information was coming through. Temperature measurements were made, and it was confirmed that Chryse is a distinctly chilly place. Even at the hottest time of the sol—around 2 p.m. local time— the temperature was no more than −22 degrees Fahrenheit, and just after dawn a thermometer would have registered more than

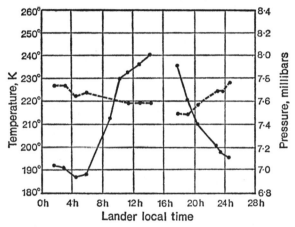

Fig. 38. Temperature (degrees Kelvin) and pressure variations on the Chryse site (Viking 1) over a typical sol.

120 degrees below zero Fahrenheit (Fig. 38). Remember, however, that Chryse is some way from the Martian equator.

Winds were light. They were strongest at about 10 a.m. local time, but even then they were no more than 14 m.p.h. breezes; later in the sol they veered, dropping to around 4 m.p.h. and coming from the south-west rather than the east. The pattern seemed to be fairly regular, sol after sol. The pink sky was cloudless, and there was no evidence that material was being blown around in dust-devil fashion. During one of our BBC television programmes, Dr. Garry Hunt even made a Martian weather forecast, based upon information sent back

from what he correctly called Man's most remote meteorological station. His forecast was: "Fine and sunny; very cold;
winds light and variable; further outlook similar." Not surprisingly, he proved to be completely accurate!

One of the most important of all the investigations was the
analysis of the atmosphere. In fact, this had already been
started during the descent of the landing capsule. Temperatures
recorded during the descent included $+27$ degrees Fahrenheit
at a height of 130 miles, but this does not imply that the upper
atmosphere is 'hot'; scientifically, temperature is measured by
the rate at which the various atoms and molecules move around,
and is not the same thing as everyday 'heat'. At 85 miles above
the ground the temperature was found to be -216 degrees

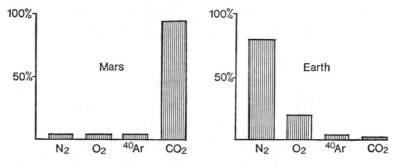

Fig. 39. Relative compositions of the atmospheres of Mars and Earth

Fahrenheit. Nitrogen was detected, which pleased those
scientists who were anxious to find life on the planet. It was
also shown that contrary to the data reported from Russia's
Mars Fleet, there was very little argon (Fig. 39).

Argon, incidentally, is of special importance. It comes in
three varieties, known as Argon-36, Argon-38 and Argon-40
(the chemical abbreviation for argon is A). The first two types
are due to the decay of radioactive potassium, while A-40 is
probably 'left over' from the very early stage of a planet's
existence, when it was still being formed out of the original
solar nebula. The Earth's atmosphere contains three hundred
times more A-40 than A-36, and five times as much A-36 as
A-38. For Mars, it was found that A-40 accounted for roughly
$1\cdot6\%$ of the total atmosphere, and that there was little of the

other varieties of argon; the ratio of A-40 to A-36 was about 3100:1. From this, it would seem that Mars must have little radioactive potassium in its crust.

The latest analyses show that the make-up of the Martian atmosphere is: 95% carbon dioxide (CO_2), 2·7% nitrogen (N_2), 0·15% oxygen (O_2), 1·6% argon, and traces of other gases—such as krypton and xenon, both of which are also found in our own air, though admittedly they are in short supply. The ground pressure was the expected 7 millibars (Fig. 38), as against an average of 960 millibars for the Earth's air at sea-level, but as the sols passed by it was found that the pressure was falling. The decrease was no more than about 0·012 millibar per sol, but it was definite, and there seemed to be a good explanation for it. Winter was approaching in the southern hemisphere, and the polar cap was starting to grow. It was suggested that the carbon dioxide condensing to form the southern cap was being withdrawn from the atmosphere in sufficient quantity to explain the falling pressure. Of course, the decline did not continue indefinitely; even before Mars passed out of range in November 1976 it had stopped.

Water vapour was also detected, but here the main information was drawn from the Orbiter. The whole question of moisture is so vital in the search for life that I propose to defer discussion of it until the next chapter, together with the analyses of the surface materials carried out by the Lander.

All in all, Viking 1 performed almost faultlessly. The only disappointment was the failure of the seismometer, or Marsquake recorder, which would—it had been hoped—register considerable ground movement; after all, there are frequent tremors on the Moon, which would be expected to be far less active than Mars. Unfortunately, the cage which had protected the seismometer during the long flight refused to shift, and all efforts to persuade it to move proved abortive. Scientists could only hope that the seismometer of Viking 2 would work; and Viking 2 was already poised for descent. It had entered Martian orbit on 7 August, and was ready to dispatch its Lander to the chosen target.

As with its predecessor, there were qualms about the site. Cydonia seemed to be unpleasantly rough. Pictures taken from the first Orbiter were not encouraging, and there was no help from Arecibo, because Cydonia is so far north on Mars that it is out of the range of Earth-based radar. After much deliberation, it was decided to abandon Cydonia and substitute Utopia Planitia, a broad plain 4600 miles north-east of the Viking 1 site, which was expected to be flattish. And it was here, on 3 September, that the second Lander touched down. The position was given as latitude +48 degrees, longitude 226 degrees west.

Events did not go as smoothly as with the first mission. After the descent capsule separated, there was a momentary power failure in the Orbiter—slight, but enough to allow the spacecraft to drift off its main communications link with Earth. Unless something could be done, all the information collected by the Lander during its descent would be lost, because the data were to be relayed immediately, and if the Orbiter could not pick up the signals all the priceless information would merely beam uselessly into space. Prompt action was called for. The planners were able to use the low-power communications link, and no data were permanently lost, though it was not until after touch-down in Utopia that the trouble was completely cured and the main communications link restored.

It was time for Mars to produce yet another of its seemingly never-ending surprises, and it duly did so. Instead of being a gently-rolling landscape, with sand-dunes and windblown dust, Utopia seemed to be of the same general aspect as Chryse. One investigator commented that it looked like 'a forest of rocks'. There were no major craters in sight—the closest large formation, Mie, lies 125 miles to the west—and there were no boulders as large as those in Chryse, while the rocks looked 'cleaner' and the general terrain was slightly flatter. On the other hand, medium and small rocks were abundant, and most of them were vesicular. Vesicles are porous holes, formed as molten rock cools at or near the surface of a lava flow, so that internal gas-bubbles escape. There is absolutely no doubt that Utopia is a volcanic landscape, and this is borne out by various other rock-types on view. There are breccias (i.e. rocks made up of fragments of other rocks welded together under the influence of high pressure and heat) and there is at least one splendid example of a xenolith—that is to say, a 'rock within a rock', probably formed when a relatively small rock was caught in the path of a lava flow and was coated with a molten envelope which subsequently solidified. There is also a great quantity of fine material similar to that in Site 1, but there are no comparable sand-dunes in sight, though there is a winding feature which looks suspiciously like the bed of a long-dry stream (Plate XII, *bottom*).

Otherwise, the results from the second Lander have not been significantly different from those of the first. The mean atmospheric pressure, 7·7 millibars, indicated an appreciable difference in level between Chryse and Utopia. Temperatures ranged between −23 and −114 degrees Fahrenheit. Winds remained light and variable, though with occasional gusts of over 30 m.p.h. The scismometer worked, to the relief of everyone, but no Marsquakes were recorded, which was a distinct surprise; if the instrument had been placed in, say, Southern California it would have recorded tremors constantly. It did, however, send back results of the tremors produced by the mechanism of the Lander itself when various operations were being carried out. The first Marsquake was recorded from Lander 2 in February 1977.

Meantime, some new manœuvres were in hand. It was known that for some weeks after 10 November there would be a loss of contact with Mars, because the planet would be too nearly behind the Sun. Superior conjunction was due on 27 November, and signals would be hopelessly blocked by the solar radio emission, after which nothing more could be expected until shortly after Christmas. Therefore, the Orbiters were ordered to do what may be called a 'general post', so that Orbiter 1 could act as the relay for Lander 2, releasing Orbiter 2 to carry out its independent programme. Lander 1 would go into what was termed its reduced phase, sending back less data every sol.

During the peak period, a Lander can contact Earth in two ways: either direct, or via the Orbiter as a relay. The quality of the transmissions is the same, but the direct link can be used for less than one and a half hours per sol (a period known as 'real time'), whereas the relay link is capable of sending back much more information much more rapidly. (All commands sent from Earth *to* the Landers are direct.) The Orbiters had been put into paths which took them round Mars in exactly one sol, so that they came back over their relevant Landers at the same time each revolution, a situation which is known as being in synchronous orbit. On 11 September this was altered. Orbiter 1 fired its engine for sixteen seconds, which was enough to break the synchronous link with its Lander and reduce the orbital period to 21·3 hours. In other words, the Orbiter now reached its lowest point every 21·3 hours instead of once every sol. Since it reached its lowest point ahead of time, the previous day's reference point would lie to the west, by a distance equal to the distance that the planet's surface can rotate in 3·3 hours (the difference between a sol and the Orbiter's revised period). Therefore, to a Martian observer it would seem that the Orbiter would be further east each day, and would appear to be 'walking' round the planet.

By 24 September the lowest point in orbit was directly over Lander 2, and another manœuvre restored the period to its synchronous value, so that Lander 2 was again fully supplied with a relay vehicle. Orbiter 2 had begun a similar 'walk', and continued to send back excellent pictures as well as a vast amount of miscellaneous information. In particular, it was ordered to study the north polar regions. By then it had become obvious that the polar caps could hold some of the vital clues in the search for life.

Chapter Eleven

THE SEARCH FOR LIFE

WE MAY BE VIRTUALLY CERTAIN that there is only one planet in the Solar System upon which liquid water can exist under natural conditions, and that planet is the Earth. Our world is unique in that it has oceans. Mars has none; the atmospheric pressure is much too low, though the presence of dry river beds indicates that surface water existed there not so very long ago by cosmical standards.

But water can be frozen in the form of ice, and if ice existed on Mars in large quantities the chances of finding life there would be increased enormously. This was agreed by everyone, and the early Mariner findings, indicating that the polar caps were made up of solid carbon dioxide, came as a distinct disappointment. Yet even before the first Mars flights, it had been thought that the caps were nothing more than wafer-thin layers of frost, containing only a very small quantity of water locked up in the frozen state. I once estimated that if all the ice could be melted at once, the water produced would be insufficient to fill a lake the depth of Windermere and the size of Wales—which shows how grotesquely wrong one can be, even when taking the best evidence available!

Probably the most startling of all the discoveries made during the first weeks of Viking research was the revelation that the caps are not, in the main, made up of solid carbon dioxide at all. Evidence accumulated that although there must be a seasonal carbon dioxide coating, the residual caps are made up of ordinary ice, and are much thicker than had been anticipated. Apart from spectrographic analysis, there was the question of temperature. During its twenty-second revolution round Mars, the Viking 2 orbiter made some temperature measurements of the north polar cap, and these seemed to be conclusive. The dark surrounding area had a range of between −36 to −27 degrees Fahrenheit at maximum, while the white area was colder, at −81 to −90 degrees Fahrenheit. This is chilly enough, but—and this is the important point—it is well above the limit at which solid carbon dioxide could persist. In fact, the cap was simply too warm, and it followed that the permanent or residual cap—that is to say, that part of the cap which is always present, and does not vanish during Martian summer—is composed of ice. Moreover, the cap is now thought to be at least half a mile thick, and the two polar caps combined may lock up from 1000 to 100,000 times as much water as now exists in the atmosphere. On the other hand, there is no carbon dioxide reservoir, and all the solid carbon dioxide coating is returned to the atmosphere when warmer weather arrives in spring.

This was a total revolution in thought, and scientists at the Jet Propulsion Laboratory went so far as to suggest that Mars might be encased in a shell of water ice, so that earlier developments could account for many of the present-day features seen above the permafrost layer.

Once this idea had taken root, the situation began to clarify. There was much more atmospheric water vapour than had been expected; indeed, above latitude +60 degrees the atmosphere was saturated. There proved to be twenty times more water vapour over the cap than over the equator. The greatest concentrations of all were in low-lying areas, such as Hellas; other regions with plentiful supplies were Amazonis and Utopia itself. Predictably, there was much less water vapour over lofty areas such as Alba Patera. Cloud phenomena also fitted in; the ground ice tended to evaporate after sunrise, producing ground fogs, and during afternoon the water vapour condensed once more as ice, so that in general terms Mars may have snowfall even though there has been no rain for a very long time. And one crater, Korolev in the north polar region, was found to be filled with ice. The first two Vikings were able to carry out detailed studies of only the northern cap, but there was no reason to doubt that the southern cap was essentially similar.

In fact, H_2O is relatively abundant on Mars. We infer that the climate is very variable over long periods, and at times it can presumably be both wet and hospitable. As a world, Mars is neither dead nor dying.

Another discovery was that there is water locked up in the rocks even in the desert regions, and this is where the Landers played the vital rôle. The first step was to analyse some of the surface material, and see what elements were present there. Each Lander was equipped with a surface sampler, known more commonly as a grab, made up of a collector head on the end of an ingeniously-constructed retractable boom composed of two ribbons of stainless steel, welded together along the edges. The head itself was, basically, a scoop with a movable lid. It was capable not only of collecting Martian 'soil' and bringing it back inside the Lander for analysis, but also of digging a trench; it was even strong enough to overturn modest rocks and sample the soil beneath. Of course, all the Lander manœuvres were first simulated on Earth to make sure that they worked properly.

There was an initial alarm with the grab of Viking 1, because a latch-pin jammed and prevented the collecting operation from being completed. The planners decided to extend the boom beyond its first position, and this proved to be successful; the obstructing pin was released, and fell to the ground, where it was subsequently photographed. (It was ironical that a three-inch pin almost wrecked one of the most complex and delicate experiments ever designed.) Finally, on the eighth sol after arrival—28 July by our calendar—the grab secured a sample, and also dug a trench three inches wide, two inches deep and six inches long. It was significant that the sides of this miniature trench did not collapse, as would have happened if the material had been very soft (Plate XIV).

The first studies of the composition of the Martian material were made by using a source of X-rays carried in the space-craft. When elements are bombarded with high-energy X-rays, the well-known phenomenon of fluorescence makes them emit X-rays of lower energy, and every element produces its own characteristic emissions, thereby betraying its identity. For once Mars provided no real shocks. It was found that there was 12% to 16% of iron, 15% to 30% of silicon, 3% to 8% of calcium, 2% to 7% of aluminium, and 0·25% to 1·5% of titanium, as well as various minor constituents. Comparisons with the Earth and Moon showed that the nearest analogy was the make-up of the lunar maria, or waterless seas. This was hardly surprising, since it had long been known that the Moon's maria are rock-strewn lava-plains.

Another interesting study concerned magnetic materials. Arrays of small magnets fixed to the body and sampler arms of Viking I were able to pick up numerous particles, and it was concluded that the soil contained 3·7% of magnetic material, a large fraction of which was probably an iron oxide called magnetite.

With a vast amount of locked-up water, an atmosphere containing a definite amount of nitrogen, and a temperature which was not impossibly low, the prospects of finding life appeared to be reasonably bright. Nothing more than very lowly organisms could be expected, but the discovery of life in any form would be immensely significant, and the experiments were put in hand as soon as was practicable.

I do not propose to go into great detail, and the account given here is badly over-simplified, but at least I hope it will serve as a guide. Basically there were three experiments, all of which were tried by the Landers. They were:

(1) *the Pyrolitic Release Experiment* (Fig. 40).—Pyrolysis is the breaking-up of organic compounds by heat. The experiment was based upon the assumption that any life on Mars would contain carbon, as is the case on Earth. One species of carbon,

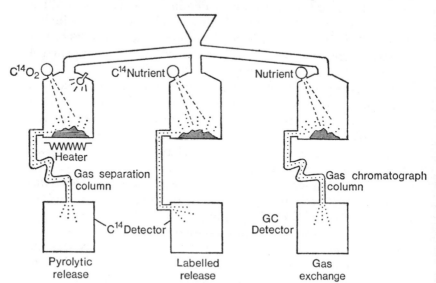

Fig. 40. The experiments involving the Viking search for life on Mars

known as carbon-14, is radioactive, so that when it is present it is relatively easy to detect. The scheme was to treat the Martian samples with this convenient carbon, and see whether they assimilated any of it.

The sample was heated in the test chamber for five days. The atmosphere in the chamber was similar to that on Mars, except that carbon dioxide and carbon monoxide, duly labelled with carbon-14, replaced part of the natural atmospheric gases. To make the situation even more realistic, the chamber was illuminated by artificial sunlight, to help in the phenomenon of photosynthesis (the production of organic compounds by carbon dioxide and water by green plants). The only major departure from actual conditions was that the artificial sunlight lacked the short-wave ultra-violet radiations which come from the real Sun, and which might well have damaged any Martian organisms present.

After an incubation period of eleven days, the chamber was heated to 1160 degrees Fahrenheit, hot enough to break up any organic compounds (pyrolysis, as we have noted). A stream of helium gas, which is inert, then flushed out the chamber, and swept the vaporized pyrolysis products into a detector capable of identifying the carbon-14 taken up by the Martian organisms —if any. Should no carbon-14 be found, it would be safe to assume that there was no biological activity or, in other words, no life; but a positive result would not be conclusive, because a 'first peak' of radioactivity in the vaporized gases could also be caused by chemical processes which did not involve living organisms. A second heating, this time to 1290 degrees Fahrenheit, should convert any trapped organic compounds into carbon dioxide, and again the labelled carbon-14 should show up, producing a second 'radioactive peak'.

Two peaks were shown, but neither was at all conclusive, and the same pattern was followed in later experiments from both the Landers. This includes samples which had been deliberately sterilized, so as to kill any organisms which might be present.

(2) *the Labelled Release Experiment.*—This also involved the useful carbon-14, and assumed that the addition of water to a Martian sample would trigger off biological processes if any organisms were present. Again the atmsophere provided was

similar to that of Mars, though in the first experiments the chamber was kept dark during the incubation period. The sample brought in by the grab was moistened with a nutrient which contained carbon-14 as well as various other substances which are taken in by Earth organisms, while the atmosphere above the sample was kept under close surveillance by a set of detectors capable of showing any traces of radioactivity. Organisms would be expected to give off gas containing carbon, and the carbon-14 would therefore betray it. Therefore, if any radioactivity in the atmosphere of the chamber were found, it would be a strong pointer to the existence of life.

The results of the preliminary experiment were startling. As soon as the first drops of nutrient were put into the chamber, the level of radioactivity rose sharply, and the excitement at Pasadena was considerable. Alas, things did not turn out to be so straightforward as had been hoped. Within a sol, the level of radioactivity dropped, and after a week it had levelled out. At the end of a seven-day period a few more drops of nutrient were put into the chamber. The immediate result was the release of more labelled carbon-14, but subsequently the total amount actually dropped by 30 per cent., after which it rose again very slowly.

The investigators simply did not know what to make of it. The behaviour of the sample did not indicate either 'life' or 'no life'; it was a complete puzzle. In a later test, a sample was 'cold-sterilized' in a way which should have made it possible to distinguish purely chemical reactions from the more sensitive biological ones. The values obtained were lower than for untreated samples, but higher than those for fully sterilized ones.

(3) *the Gas Exchange Experiment.*—This was expected to be the most sensitive of all the tests, and for once carbon-14 was not involved. The main assumption was that any biological activity on Mars would involve the presence of water, and the idea was to see whether providing a sample with suitable nutrient would persuade any organisms to release gases, thereby altering the composition of the artificial atmosphere inside the test chamber. This atmosphere was made up of carbon dioxide, together with the inert gas krypton to act as a calibration standard, and helium to bring the pressure up to an acceptable

level. The sample was held in a porous cup above the chamber floor, and a rich mixture of common nutrients in water solution was added to the bottom of the chamber—a mixture which was generally nicknamed 'chicken soup', though I doubt whether any gourmet would have found it palatable! For the first week the level of liquid was kept below the cup, so that only water vapour could be transferred to the sample. After that, the level of the liquid was raised so that the sample was actually wetted. The atmosphere in the chamber was regularly tested to see if any changes were occurring.

Again there was an immediate surprise. As soon as the so-called chicken soup was put into the chamber, there was a violent release of oxygen from the sample—fifteen times as great as could be explained by any known process. Carbon dioxide was also released, together with a certain amount of nitrogen, but after a while there was a noticeable levelling-off, and when the soil was wetted no more oxygen was set free; the carbon dioxide content decreased. The investigators more than two hundred million miles away frowned thoughtfully. Nothing seemed to make sense, and even the fact that the sample came from a desert, thereby being unused to appreciable moisture, did not help much in finding an explanation. Eventually they came round to thinking that the reactions were chemical rather than biological, due either to the release of oxygen stored in the sample or else by the reaction of some unstable oxidant with the nutrient.

I have dwelt at some length upon these experiments because they were so important and so novel. Trenches were dug all round the two space-craft, and as many samples as possible were examined (it was commented that the areas round the Landers began to look rather like sites for mining operations!) and samples were also obtained from beneath rocks which were overturned or pushed aside by the grabs. One rock, tackled by Viking 2 in October, obstinately refused to budge, so that evidently most of it lay buried beneath the surface, but a smaller one was pushed a total of eight inches, and on 25 October, at 6.45 a.m. Martian time, a scoopful of soil was obtained from a third rock which had been shifted by the grab half an hour earlier. Yet with these samples, too, the results remained inconclusive. The situation was summed up very neatly by Dr.

Gerald Soffen, one of the principal investigators, after some of the results from Lander 2. "All the signs suggest that life exists on Mars," he said wryly, "but we can't find any bodies."

I think that most people, myself included, expected the Landers to give a final answer to the question which had been tantalizing astronomers for generations; but this did not happen, and all we can do is to draw some general conclusions. We must remember that everything we have found out is based upon small quantities of material taken from only two sites on Mars. (The amount of 'soil' used in the various experiments would hardly fill an ordinary test-tube.) It is just possible that the choice of sites was unfortunate, and that neither Chryse nor Utopia is typical of the surface as a whole, but this I rather doubt—despite the earlier bad luck when Mariners 6 and 7 flew past Mars but totally missed the volcanoes and the deep chasms. Martian chemistry may well be stranger than we anticipated, so that the curious results of the experiments do not indicate any biological activity. Or it may be that there really are organisms, behaving in a manner which is totally un-expected.

The one thing about which we can be positive is that advanced life on Mars is out of the question at the present time. Whether it has ever existed, or whether it will evolve in the future (even disregarding possible interference from Earth) remains to be seen, but in our own era we are limited to tiny organisms at best. So far there is no evidence of any life, but we cannot state categorically that it does not exist—even though we have come a long way since the days when Percival Lowell wrote so enthusiastically about his canal-building Martians.

Chapter Twelve

MARS—AFTER VIKING

There must be rivers on Mars. The mere existence of continents and oceans proves the action of forces of upheaval and of depression. There must be volcanic eruptions and earthquakes, modelling and remodelling the crust. Thus there must be mountains and hills, valleys and ravines, watersheds and watercourses . . .

THIS IS NOT A REPORT from the Jet Propulsion Laboratory. It is a quote from R. A. Proctor, published as long ago as 1871. Of course Proctor was wrong; but had he written in the past tense, and begun "There must have been rivers on Mars . . . " he might have been close to the truth. It is at least possible that in our own era we are seeing Mars at its very worst.

To my mind, the most fascinating of all the features shown by Mariner 9 and the Vikings are the so-called river beds and dry streams. We have no final proof that they were water-cut, but all the evidence indicates that they were. There can be no doubt whatsoever that they were produced by flowing liquid (one has only to look at them) and very runny lava is the only possible alternative; but to believe that lava of any kind could produce the effects we observe is difficult to accept. Moreover, the discovery that the residual polar caps are of water ice, and that there is a great deal of H_2O locked up in the surface materials, is a further strong pointer.

Note, also, that the river beds (as I propose to call them) are neither badly eroded nor filled up with dust. This applies also to other features, and the volcanoes show no signs of having been worn away. If the river beds were extremely ancient by geological standards—that is to say, millions of years old—they would hardly have been preserved to the extent that they actually are. Erosion would have taken its toll, particularly since we know that dust-storms and sand-storms are common enough, and that both sand and dust are highly abrasive even in an atmosphere as thin as that of Mars. Not that I am suggesting for one moment that the river beds are young by everyday standards, and the fact that some of them have craters inside them shows that the water stopped flowing long ago, but in all probability we are reckoning in tens of thousands of years rather than millions.

There is a paradox here, obvious enough when pointed out. The low atmospheric pressure—nowhere as high as 10 millibars, and therefore corresponding to what we normally call a laboratory vacuum—means that liquid water cannot exist on Mars today. Therefore, in the days when the rivers and the streams ran, there must have been a much denser atmosphere than there is now, and we cannot go back too far in time, though I agree that it is always dangerous to give anything in the nature of a precise time-scale when the data are so scanty.

Neither does it seem that the present atmosphere can be merely the lingering remnant of an original atmospheric mantle. If so, then there should be very little argon-40 as compared with other varieties of argon. All of which raises the intriguing possibility that conditions on Mars may change dramatically and perhaps cyclically, so that there are what we may call 'fertile' periods, with relatively thick atmosphere and plenty of surface water, alternating with near-sterile ice ages such as that which prevails at the moment.

Everything hinges upon the amount of atmosphere, and if these ideas are basically correct we have to explain just how an atmosphere can appear and vanish again with reasonable regularity. Two theories have been proposed. Either may contain something of the truth; both may be completely wrong, but at any rate they are worth discussing.

The first—and, I must add, the more popular—was outlined in 1971 by Carl Sagan, the American astronomer who has always maintained that life of a kind may survive on Mars, and who is the founder of a new science known as 'exobiology'. The name is self-explanatory, and although the theory is as yet founded upon little more than speculation it may well become highly important in the future. The Sagan explanation depends upon the changing axial tilt of Mars.

As we have noted, our Earth is subject to an effect known as precession. Because of the pulls of the Moon, Sun and (to a very slight degree) the planets upon the Earth's equatorial bulge, the direction of the axis of rotation is not absolutely constant. The effect is much the same as that of a gyroscope which is running down and is starting to topple, the main difference being that the axis of the gyroscope completes a circle in a few seconds, while the axis of the Earth takes 25,800 years. This, of course, alters the positions of the celestial poles. At present the axis points northward to a position in the sky near the star Polaris in Ursa Minor, the Little Bear, which seems to remain almost motionless, with the entire sky revolving around it once a day. (There is no bright south polar star; the nearest candidate is Sigma Octantis, which is frankly a feeble substitute, and is none too easy to see with the naked eye.) In the days of the Egyptian Pyramid-builders, the northern pole star was Thuban in the constellation of the Dragon; in 12,000 years from now it will be the brilliant blue star Vega. Incidentally, precession also changes the position of the equinox, the point at which the celestial equator cuts the ecliptic (the apparent yearly path of the Sun in the sky). In early historic times the spring equinox lay in Aries, the Ram; by now it has been shifted into the adjacent constellation of Pisces, the Fishes, though it is still known as the First Point of Aries.

The angle of inclination of the Earth's axis does not alter very much, even though many authorities believe that it has a marked effect upon our climate over the ages, and of course the Earth's orbit is not far from circular; the fact that we are slightly closer to the Sun in December than in June makes very little difference to the seasons. Not so with Mars, where the orbit is more eccentric, and the precessional period is of the order of 50,000 years. The Martian axial inclination ranges

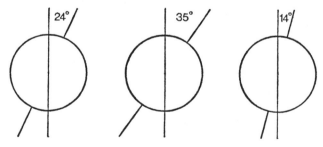

Fig. 41. The range of inclination of the axis of Mars. The present value is 24°, very similar to that of the Earth

between 35 degrees and only 14 degrees, so that it is sheer coincidence that the present value (24 degrees) is almost exactly the same as ours (Fig. 41).

When Sagan described his theory, in 1971, it was believed that the whole of the polar caps of Mars was made of carbon dioxide, and that the southern cap vanished completely during summer in that hemisphere, whereas the northern cap did not. This, he wrote, was because when Mars is at perihelion the northern hemisphere is tilted away from the Sun, and the heat at the pole is never sufficient to vaporize the cap completely. In 25,000 years' time—half a precessional cycle—the situation will be reversed; the southern cap will be turned away from the Sun at perihelion. Midway through the precessional cycle, roughly equal amounts of cap material will exist at each pole, but it will be conceivable that both caps will totally disappear at some stage of the Martian year, so that the volatiles released will thicken the atmosphere enough to change the entire climate. Also, the maximum tilt of 35 degrees will undoubtedly mean that the pole facing the Sun at perihelion will be warmed to a considerable extent.

It all seemed very plausible—and in modified form it still does, even though we now know that the main caps are of ordinary ice, and are much too thick to be completely melted under present conditions. Obviously an atmosphere produced in such a way would be temporary. When the axial inclination returned to anything like the value that it has today, ice age conditions would prevail once more, and any developing life-forms would have their evolution rudely suspended.

One difficulty to be faced is that we must find some way of explaining how a large amount of carbon dioxide can be put back into the atmosphere. Extra carbon dioxide would increase the opacity of the atmosphere as well as the temperature of the surface, and it seems that liquid water cannot run unless the ground pressure goes up to at least 100 millibars. The revelation that the residual north polar cap is made up of ice sets us a real problem, because unless there is a great deal of carbon dioxide bound up in the main surface of the planet the CO_2 content of the atmosphere—and therefore the pressure—may not rise much in the future.

The same objection can be held against Theory No. 2,

according to which Mars goes through spells of intense vulcan-ism, when tremendous quantities of gases (including water vapour) are sent out from beneath the crust, and the atmosphere is accordingly thickened for a while. Another trouble here is in deciding just why this should happen, though it is conceivable that activity 'builds up' over a very long period and results in a sudden series of outbursts. All in all, it is still premature to say definitely that the climates of Mars change in a cyclic manner, though the river beds indicate that things were much less unpleasant in the past than they are now.

There is no doubt that the surface represents a volcanic landscape. Quite apart from the obvious shield volcanoes, we have overwhelming proof from the vesicular rocks seen at both Lander sites. But what about the craters? Here there are two sharply-opposed schools of thought: either the craters are of volcanic origin (using the term in its broadest sense) or else they are due to the impacts of meteorites. This is also true of the craters of the Moon, and I feel that I must say a little about the whole problem, though I do not pretend to be unprejudiced, and I am well aware that my views will be hotly criticized. Perhaps I should say at the outset that both theories may be valid to some extent!

The craters of the Moon are scattered all over the lunar sur-face. They range from vast enclosures well over a hundred miles in diameter down to tiny pits, and many of them have terraced walls and high central mountains. In profile, however, a lunar crater is more like a shallow saucer than a deep mine-shaft; for instance, Theophilus, one of the most prominent for-mations on the Moon, is 64 miles in diameter, but the crest of its rampart is a mere $3\frac{1}{2}$ miles above the level of the floor, and the floor itself is depressed well below the outer surface. Larger walled plains, many of them devoid of central elevations, are even shallower in proportion. There are no shield volcanoes of Martian calibre, and the chief mountain ranges appear to be nothing more than the walls of the circular seas or maria; for example, the lunar Apennines make up part of the border of the huge, well-defined Mare Imbrium or Sea of Showers.

According to the impact theory, the circular structures—walled plains, maria and all the large craters—were produced by a kind of cosmical bombardment in past ages. Impact craters must certainly exist, and after all they occur also on the Earth, the best-known example being the Meteor Crater in Arizona, which is almost a mile wide and which was definitely formed by a missile which hit the desert in prehistoric times. On the other hand a random bombardment would be expected to strew craters haphazardly all over the Moon, and this is not what we find. There are definite laws of distribution. The walled plains tend to line up, even when the members of a line are of palpably different ages, and when one crater breaks into another (as happens frequently) it is an almost invariable rule that it is the smaller formation which intrudes into the larger, not vice versa. The central peaks of craters are often crowned by summit pits which look very like volcanic structures. The samples brought home from the Moon by the Apollo astronauts and the automatic Russian probes are of volcanic nature, and there is little evidence of meteoritic material. In fact one famous expert on the subject, J. G. McCall, has asked plaintively "Where have all the meteorites gone?"

Again according to the impact theory, the main circular maria were produced around four thousand million years ago (or four æons; one æon is taken to be a thousand million years). Tremendous lava-flows then poured forth, flooding the impact-produced basins; this went on until just over three æons ago, after which it stopped rather suddenly. Meanwhile, smaller meteorite falls had produced craters and walled plains, some of which were flooded with Mare-material while others were not. The main cratering period thus ended a very long time in the past, and even the youngest of the major craters of the Moon—such as Tycho, which is the focal point of a system of bright streaks or rays—are held to be more than one æon old. Impact pits formed since then are relatively small. I mention this problem of age because it is extremely important when we come to think about Mars.

For various reasons, including the laws of distribution, I have very little faith in the impact theory as a whole. Nobody will question the existence of meteorite pits, but I very much doubt whether the larger structures can be explained by bombardment. This leads one on to a volcanic hypothesis, and it has been suggested that the main craters are of the caldera type—a caldera being a volcanic crater formed by the collapse of the surface into an underground cavity, either by the rapid eruption of large quantities of molten magma or else by the withdrawal of magma from a magma-chamber. There is a caldera atop the Earth's Vesuvius, for example, and of course calderas are prominent upon the Martian giants such as Olympus Mons.

There are other plausible theories, such as that described by Dr. Allan Mills of Leicester, who introduces the idea of a fluidized bed. If a stream of gas flows beneath a particular material, and the flow is then increased, the bed will expand; bubbles will appear in it, and finally material will be blown out of the vent, leaving craters which in form are strikingly like those of the Moon. I do not propose to go more deeply into the matter here, if only because I have done so elsewhere,* and I will do little more than mention the craters on other worlds. Results from the two-planet probe Mariner 10 show that Mercury has a surface of lunar type, with craters of the same kind, while radar measurements from Earth show that there are many large, rather shallow craters on cloud-covered Venus. The craters on Venus can hardly be due to impact, because it would need a very large meteorite to survive a complete fall through that dense atmosphere without being destroyed, and there must also be considerable surface erosion, so that the craters there are presumably rather young. So far as Mercury is concerned—well, we may regard it as a kind of second Moon with respect to its surface features.

Now let us return to Mars. The craters there are not identical in type with those of the Moon or Mercury, but of course the conditions are different, because Mars has an atmosphere. Generally speaking, the laws of distribution follow the same kind of pattern, with small craters breaking into larger ones; some of the formations have prominent central peaks, while others have none. According to many authorities, impact has been the main cratering process, and it is of course true that Mars is closer to the asteroid belt than we are, so that what we may call cosmical débris would be expected. Basins such as Hellas and Argyre are also attributed to impact.

Personally, I have the gravest doubts. As with the Moon, I would not for one moment deny the existence of meteorite pits, but with large craters the situation is different. Bearing in

* *Guide to the Moon*, Lutterworth Press, 1976.

mind the near-certainty of liquid water having existed on Mars within the last million years, it would be illogical to suggest that the craters can be anything like as ancient as those of the Moon, because of erosion. Therefore, on the impact hypothesis we must assume that a tremendous bombardment took place in the geologically recent past, which does not make sense. If Mars had been scarred in this way, the Earth could scarcely have escaped unscathed—unless, of course, Mars was unfortunate enough to encounter a concentrated swarm of large meteoroids, which would be too much of a coincidence.

There seems no point in saying more as yet. If astronomers cannot agree about the origin of the craters of the Moon—when they can not only study photographs taken from close range, but even analyse samples of the surface material—it is hardly likely that they will be more in accord when it comes to Mars. At the moment the two schools of thought are firmly entrenched, and only time will show which is more nearly right.

It is interesting to look back once more, and note how our ideas about Mars have seesawed. To Herschel, and to Lowell, it was a world capable of supporting life; Lowell believed in a civilization compared with which *homo sapiens* would seem primitive. In the years after Lowell's death, Mars became a dying planet, devoid of inhabitants and even of advanced vegetation. Then, following Mariners 6 and 7, it was dismissed as a cratered waste, dead and sterile. It has taken Mariner 9 and, above all, the Vikings to prove that we are dealing with a world which may be going through nothing more than a temporary ice age. There are many problems to be solved, but with each new puzzle the fascination of Mars increases.

Chapter Thirteen

PHOBOS AND DEIMOS

MOST PEOPLE, I suppose, have read Jonathan Swift's classic *Gulliver's Voyages*. To be more precise, most people have read the voyages to Lilliput (the country of the midgets) and Brobdignag (the country of the giants). The remaining Voyages are less famous, but one of them is particularly relevant in a discussion of Mars. Dr. Lemuel Gulliver is said to have visited Laputa, an airborne island which probably qualifies as the first fictional flying saucer. The Laputan astronomers were so skilful that they had discovered "two lesser stars, or satellites, which revolve about Mars, whereof the innermost is distant from the centre of the primary planet exactly three of his diameters, the outermost five; the former revolves in the space of ten hours, the latter in twenty-one and a half". In other words, the inner satellite moves round Mars so quickly that it completes its circuit in less than a sol.

The Voyage to Laputa was written in 1727. At that time there was no telescope in existence powerful enough to show the two satellites of Mars that we now know to be genuine, and this was also true when Swift died in 1745. Five years later, the French novelist Voltaire wrote a rather strange story, *Micromégas*, in which the Solar System is visited by a being from the star Sirius. Voltaire also credited Mars with two moons.

In fact, the Martian satellites were not discovered until 1877, and the Swift and Voltaire stories have led to some peculiar speculations. It has even been suggested that our remote ancestors had optical instruments which enabled them to track the satellites down. Unfortunately for this intriguing idea, the true explanation is very simple, and Voltaire himself wrote it down. He pointed out that because Mars is further away from the Sun than we are, how can it possibly manage with less than two moons?

There was also something of a progression in the numbers of planetary satellites known in the mid-eighteenth century. Mercury and Venus appeared to be unattended. (They still

167

are. A satellite of Venus has been reported now and then, but is nothing more than a telescopic 'ghost'.) The Earth, of course, had one moon. Jupiter had four known satellites, all discovered by the earliest telescopic observers; Galileo saw them in 1610, and others detected them at about the same time. Saturn had a retinue of five: Titan, discovered by Christiaan Huygens in 1655, and Iapetus, Rhea, Dione and Tethys, all found by Giovanni Cassini between 1671 and 1684. So there was the progression: Venus 0, Earth 1, Jupiter 4, Saturn 5. It was logical to give two attendants to Mars, and this is what both Swift and Voltaire proceeded to do.

As larger telescopes were built, more planetary satellites came to light. By 1850 the grand total was eighteen—Earth 1, Jupiter 4, Saturn 8, Uranus 4 and Neptune 1. Yet only the Earth among the inner group of planets seemed to be accompanied. In 1783 William Herschel made an unsuccessful search for a Martian satellite, and in 1862 and 1864 Heinrich d'Arrest, at the Copenhagen Observatory, was equally luckless. The general feeling among astronomers was that the poet Tennyson was right in describing 'the snowy poles of moonless Mars'.

Then came the close opposition of 1877, when Schiaparelli drew his map of Mars and described the canal network for the first time. Over in the United States, a well-known observer named Asaph Hall decided to renew the attempt to find a satellite. He was well equipped for the search, since he was able to use the 26-inch refractor at Washington—one of the largest telescopes in the world at that time, and also one of the best (the object-glass was made by Clark, whose skill was second to none). Hall began work in early August. At first he was as unsuccessful as Herschel and d'Arrest had been, and apparently he was on the verge of giving up when his wife persuaded him to continue for at least another few nights.

On 10 August he began observing as usual. For some time he saw nothing unusual about the background of stars, but at 2.30 in the morning of 11 August he caught sight of a very faint object close to Mars which seemed to be much more promising.

Unfortunately, fog rising from the nearby Potomac River came up before he had had time to do more than make a quick observation of the suspected object, and the next four nights were useless, as cloud and mist prevailed. Finally, on 16 August, the weather cleared. Hall was able to recover his suspected satellite, and he saw that it was moving along together with Mars, so that it was a true attendant. On the following night there were startling developments. The original satellite was seen again, and another was found, even closer-in to Mars.

Hall's announcement, made on 18 August, caused a great deal of interest, which increased when it became clear that the inner satellite at least was a most remarkable object. In Hall's own words, written a few days later: "At first I thought that there were two or three moons, since it seemed to me at that time very improbable that a satellite should revolve around its primary in less time than that in which the primary rotates. To decide this point, I watched the moon throughout the nights of 20 and 21 August, and saw that there was in fact but one inner moon, which made its revolution around the primary in less than one-third the time of the primary's rotation, a case unique in the Solar System."

Certainly this was very strange. The inner moon—actually the second in order of discovery—proved to have a revolution period of a mere 7 hours 39 minutes. Therefore, reckoning by this satellite, the Martian month was shorter than the sol. Swift had been right, even though his description had been no more than a shot in the dark.

Hall, as discoverer, had the honour of naming the satellites. He wrote: "Of the various names that have been proposed . . . I have chosen those suggested by Mr. Madan of Eton, England, viz. Deimos for the outer satellite: Phobos for the inner satellite. These are generally the names of the horses that drew the chariot of Mars; but in the lines referred to (in the fifteenth Book of Homer's *Iliad*) they are personified by Homer, and mean the attendants, or sons of Mars . . . 'He (Mars) spake, and summoned Fear (Phobos) and Flight (Deimos) to yoke his steeds'."

Phobos and Deimos they have remained. Almost a century later, Mariner 9 photographed craters on each. The main crater on Phobos has been named Stickney, because the maiden name of Hall's wife was Angeline Stickney. And the two most prominent craters on Deimos have been named, appropriately, Swift and Voltaire.

Encouraged by Hall's success, other astronomers set to work

F*   169

to detect new satellites. They failed. In much more recent times a search was carried out by G. P. Kuiper, using the 82-inch reflector at the McDonald Observatory in Texas during the oppositions of 1952, 1954 and 1956. He concluded that no satellite larger than one mile in diameter could exist, and I think we may now be sure that Phobos and Deimos are the only natural attendants (Plates XV and XVI).

Both are tiny, which explains why they were not discovered until 1877. Moreover, both are irregular in shape; Phobos has a longest diameter of 14 miles, Deimos about 8½. From Earth they look like tiny specks of light, and before the voyage of Mariner 9 we knew nothing about their surface features. The stellar magnitudes have been given as 12·1 from Phobos and 13·3 for Deimos. This is not particularly faint, but the satellites are so close to Mars that they are drowned in light; Phobos can never be more than 20 seconds of arc away from the Martian limb, and Deimos 65 seconds, which is not very much. With a moderate telescope, the only way to see them is to block out the disk of Mars with a device known as an occulting bar. I have managed to see both satellites with my 15-inch reflector, but only under exceptionally good conditions, and even then with some difficulty. Antoniadi claimed that Phobos was white and Deimos bluish, but I fear that he was drawing upon his imagination. To see colour in a spot of light as dim as Deimos is, frankly, impossible.

It is plain that Phobos and Deimos are quite unlike our own massive Moon. Their escape velocities are low—around 30 m.p.h. for Phobos and 15 m.p.h. for Deimos—so that bringing in a space-ship would be more in the nature of a docking operation than a conventional landing, and an astronaut standing on the surface of either satellite would have practically no feeling of weight.

Phobos moves in an almost circular orbit, at a distance of 5800 miles from the centre of Mars, and in the plane of the planet's equator. This means that it is no more than 3700 miles above the Martian surface, which is roughly the same distance as that between London and Aden. An observer standing at a high latitude on Mars would never be able to see Phobos at all, since it would remain below the horizon from latitudes greater than 69 degrees north or south. (Note, though, that it can be seen from both the first two Viking sites, since Chryse and Utopia are nearer the equator than this limit.)

Antoniadi worked out what our hypothetical Martian observer would see, and he was to all intents and purposes right. From the equator of Mars, Phobos would rise in the west, cross the sky in only $4\frac{1}{4}$ hours, and set in the east, during which time it would go through more than half its cycle of phases from new to full. The interval between successive risings would be 11 hours 6 minutes, and Phobos would seem appreciably larger when high up than when low down. Often it would be eclipsed by the shadow of Mars—in fact, about 1330 times every Martian year, and it would be shadow-free for its whole passage across the sky only for restricted periods near midsummer and midwinter. Even then it would be fairly useless as a source of illumination, and would shed no more light upon Mars than Venus does to us.

Deimos, moving at a distance of about 12,500 miles above the surface of Mars—that is to say, about the distance of Australia from England—has a revolution period of 30 hours 18 minutes. It, too, has an orbit which is virtually circular and in the equatorial plane (Fig. 42). As Mars spins, Deimos almost keeps pace with it, so that it remains above the horizon of the

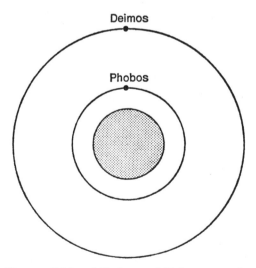

Fig. 42. Orbits of Phobos and Deimos, to scale

Martian equator for sixty hours consecutively, during which it passes through its phase-cycle twice. It too is invisible from the polar regions; any observer above latitude 82 degrees north or south will never see it. It would be eclipsed 130 times every year, but in any case its phases would be none too easy to see with the naked eye. It would look rather like Venus as seen from Earth, though rather larger and appreciably dimmer.

The two moonlets undergo all sorts of eclipse and occulation phenomena, rather as though playing cosmic hide-and-seek. They can, of course, pass in front of the Sun, but they are too small to blot out the disk, so that total solar eclipses can never occur. Phobos can hide one-third of the Sun, and passes across the disk in 19 seconds; this happens 1300 times a year. Deimos makes 120 crossings, but covers only one-ninth of the disk, and takes two minutes to pass right over. Future observers will no doubt enjoy watching the spectacle of both satellites silhouetted against the Sun at the same time, as can happen quite often. Also, Deimos can be eclipsed by Phobos.

I cannot resist mentioning a peculiar theory proposed in 1959 by Iosif Shklovsky, who is undoubtedly one of the world's great astronomers,* but whose views about extraterrestrial life are, perhaps, a little extreme. There had been suggestions that Phobos was speeding up in its path, so that it was gradually spiralling downward and might be expected to crash-land on Mars in from 35 to 40 million years from now. The calculations were made originally by B. P. Sharpless in 1945, and had been supported by various other astronomers. It was also noted that although Phobos was outside the Roche limit for Mars—that is to say the 'danger zone', inside which a fragile body would be disrupted by the gravitational pull of the planet—it was not very far outside.

Shklovsky caused something of a sensation by his claim that this speeding-up was due to friction against the Martian atmosphere. He went on to point out that at a height of more than 3000 miles above the surface, this atmosphere must be very thin indeed, so that it would be unable to influence a moving satellite of appreciable mass. Therefore, reasoned

* Shklovsky's great achievements have been in the realm of astrophysics, notably in connection with the synchrotron radiation from the Crab Nebula.

Shklovsky, the mass of Phobos must be negligible. It follows that it must be hollow, and so presumably of artificial construction —in fact a space-station, launched by the Martians for reasons of their own! The idea was eagerly seized upon by flying saucer enthusiasts, and it was even suggested that Herschel and d'Arrest had failed to find Phobos (or Deimos) for the excellent reason that they had not been sent up from Mars much before 1877 . . .

Astronomers in general were not impressed, and one has the feeling that most members of the Soviet Academy of Sciences were somewhat embarrassed. Efforts were made to account for the acceleration of Phobos by theories which did not involve space-stations. Finally, with the Mariner photographs, the idea of an artifical Phobos was quietly forgotten. The whole episode had its amusing side, and was illuminating inasmuch as it showed that even after the start of the Space Age the idea of intelligent Martians—even extinct ones—had not been entirely ruled out in some quarters.

Until 1969 our ignorance about the physical characteristics of Phobos and Deimos was complete. The first direct information came from Mariner 7. A photograph taken during the fly-by showed Phobos silhouetted against the disk of Mars, and it was seen that the satellite was extremely dark; also, it was irregular in form, and seemed to be oval rather than spherical. Mariner 9 took pictures from close range, and showed that Phobos is shaped rather like a potato. Deimos proved to be equally irregular, and both satellites gave the impression of being lumps of rocky material, possibly the fragments of a larger body which had been broken up in some manner. Also, both were pitted with craters.

Further photographs of the satellites have been taken from the Vikings, and several interesting points have emerged. First, both Phobos and Deimos have rotations which are captured or synchronous; that is to say, their periods of revolution round Mars are equal to their axial rotation periods—approximately $7\frac{1}{2}$ hours for Phobos, $30\frac{1}{2}$ hours for Deimos. This means that each satellite will keep the same face turned toward Mars all the time. An observer on the Martian surface would never be able to see the 'far side' of either moonlet.

There is nothing surprising in this, and in fact the Moon

behaves in the same way with respect to the Earth. Until 1959, when the Russians sent a camera-carrying probe right round the lunar globe, we knew nothing positive about the averted hemisphere. Tidal friction over the ages has been responsible for this state of affairs, and the same applies to both the satellites of Mars, which have, so to speak, been 'braked'—though let it be added that the satellites do rotate with respect to the Sun, so that all parts of their surfaces are bathed in sunlight at some time or other. Also, the longer axes of the satellites are directed toward Mars. On Phobos, the sub-Mars point, i.e. the point on the surface of Phobos from which Mars would appear directly overhead, is fairly closely outside the wall of the crater which has been named Stickney, while on Deimos the analogous point is not very far from Swift and Voltaire.

Inevitably, the origin of the craters on Phobos and Deimos has caused argument. Volcanic or meteoritic? Internally produced, or the result of bombardment? On Phobos, the largest crater—Stickney—is five miles in diameter, which is getting on for half the diameter of Phobos itself. Calculations

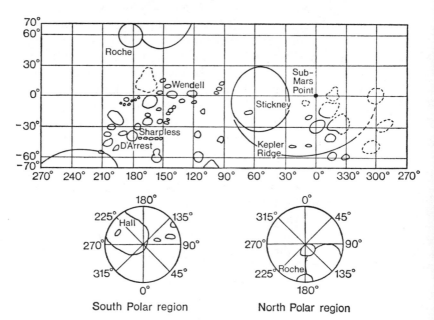

Fig. 43. Map of Phobos

have shown that if it had been formed by the impact of a meteorite, the energy produced would have been equivalent to about 100,000 atom bombs of the type used in World War II. Unless Phobos is remarkably rigid in structure, it would seem that bombardment of this type would be close to breaking the satellite up completely. And Stickney is not the only large crater; another has been named, very properly, in honour of Asaph Hall (Fig. 43).

It is also noticeable that, as with the Moon and other crater-scarred bodies, the arrangement of the surface features is not random. One particularly good picture of Phobos, taken from Viking 1, shows that the usual law of 'smaller breaking larger' applies; there are pairs and strings of craterlets, and long, narrow depressions or striations, some of which look suspiciously like strings of pits. It could well be that the craters are of internal origin: blowholes, perhaps. Up to now (1977) the smallest craters recorded on the satellites are a mere 130 feet across.

Phobos and Deimos have blackish surfaces. As reflectors, they are even less efficient than our Moon, whose overall albedo is a mere 7 per cent. It has even been said that Phobos and Deimos are the darkest bodies in the Solar System, though it would be premature to make any definite claim simply because our information is so scanty. However, the similarity between their albedos and that of basaltic lava may be more than coincidental, and it seems, too, that each satellite is covered with a thin layer of dark dust. The first real proof of this came from Mariner 9. As we have noted, the satellites are regularly eclipsed by the shadow of Mars, and when this happens the surface temperature of the eclipsed moonlet drops sharply, because the supply of direct sunlight is temporarily cut off. Mariner 9 carried a device called an infra-red radiometer, capable of measuring temperatures very accurately. The rate at which Phobos warmed up as it came out of the shadow was a key to the depth of the insulating layer on its surface, which proved to be no more than a few millimetres thick. Some Earth-based observations carried out from America by B. Zellner, for Deimos, gave similar results. Zellner was unable to measure the temperature changes, but he obtained polarization measurements of Deimos which indicated that the surface was dust-coated rather than bare rock.

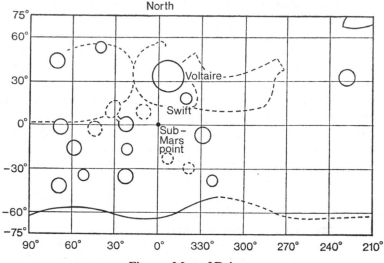

Fig. 44. Map of Deimos

We have no definite information about the origins of the satellites. It is tempting to suggest that Phobos and Deimos may be captured asteroids, and the idea is plausible at first sight; Mars is much closer to the main belt than we are, and no doubt there are many 'Mars-grazers'. The trouble is that the orbits of both the satellites are practically circular and in the plane of the Martian equator, whereas a captured asteroid would be expected to have a very elliptical orbit with high inclination. This may not be an insuperable objection, but it is fair to say that the present orbits do not seem to be what would be predicted on the capture theory.

There is no chance that the satellites were flung out of Mars itself. It has been suggested that they were produced from a ring of material which used to encircle Mars, but there are objections to this also. Possibly both Phobos and Deimos are the fragments of a larger body or bodies which broke up for some reason or other. But as yet we have to admit that we do not know, and we must await further information. We may not be able to decide one way or the other until we can obtain samples from the satellites and analyse them in our laboratories. At any rate, it is generally agreed that Phobos and Deimos are likely to be very old, and their ages may be as great as those of

176

the Earth and Mars. Certainly they give no impression of being youthful.

If Mars once had surface water (and of this there seems little doubt), tidal effects will have been feeble. The tides due to the Sun will have been weaker than on Earth, because Mars is smaller and more remote, while no measurable tides could have been produced by either Phobos or Deimos.

Tiny though they are, the moonlets of Mars are intriguing little bodies, and they may have much to tell us—even though we must regretfully abandon the idea that they could be made up of metal shells rather than conventional rock!

Chapter Fourteen

MARS IN THE FUTURE

MARS HAS PROVED to be a world of surprises. It is not in the least the kind of planet we expected it to be, but despite the regrettable absence of Martians it would be wrong to infer that what we have found has been disappointing. In many ways Mars is not particularly unfriendly, and there is at least a chance that in the far future conditions there will improve.

Of course, this is an academic point only, because the climate will not change much for many thousands of years, and in our own era we must put up with Mars as it is—cold, lacking in surface water, and with a painfully thin atmosphere composed of gases which we could never breathe. The best we can say is that there seems to be nothing which will prevent manned space-ships from landing undamaged.

Obviously there is an immediate need for more Viking-type probes. As yet we have direct information from only two sites, Chryse and Utopia, neither of which may be completely typical of the rest of the planet—any more than, say, the Desert of Gobi or the Siberian tundra would be representative of the Earth as a whole (though I agree that Mars is rather more likely to be uniform). Moreover, we have not solved the tantalizing problem of the existence or non-existence of life.

There is bound to be something of a parallel with the progress of the exploration of the Moon. The lunar venture started with the dispatch of a crash-lander; then came orbiters, soft-landers, and then manned expeditions, coupled with automatic 'rovers' such as the Russian Lunokhods, and with unpiloted vehicles which could bring back samples. The most important difference is in the time-scale. Luna 2 made the first direct contact with the Moon in 1959, and only a little more than ten years later Neil Armstrong stepped out on to the Sea of Tranquillity. It is already more than ten years since the first successful probe flew past Mars, but it would be hopelessly optimistic to expect a manned trip yet awhile.

Quite apart from future Vikings, which could land in various parts of the planet and act as transmitting stations, there are two prospects which are distinctly exciting and within our capability. One is that of a Martian Rover, which will be sent to the surface and will then trek around, controlled from Earth. There seems every chance that the Russians will attempt this, probably before they send any cosmonauts to the Moon. Obviously a Martian Rover would be more complex than a Lunokhod, mainly because of the greater distance involved; a command to a Lunokhod will reach its target in a second and a quarter, but for Mars the delay-time must always be several minutes. Also, more power will be needed, not only for moving the Rover itself but also for transmission of the data collected.

Areas such as Chryse and Utopia are extremely rocky, and other parts of Mars are unlikely to be smoother, which means that 'driving' will be a real problem, though fortunately it does not seem likely that there will be dangerous dust-drifts. There is also the point that Vikings may be unable to come down in some of the really fascinating areas, while Rovers could possibly reach them. It might even be practicable to explore the cap zones in this way.

The second practicable scheme is to send an automatic probe, land it on Mars and then bring it home again, carrying samples of the surface material. There is no need to stress how important this would be, and it is indeed an essential part of the overall programme. Arguments are now going on between scientists who want to concentrate upon Rovers, and those who are anxious to handle samples of the Martian crust. Eventually it may well be that the two projects are attempted at about the same time. If I had to make a guess, I would say that they will have been achieved before 1990. So far as returned samples are concerned, they will no doubt be given a preliminary examination on board an Earth orbiting space-station before being brought down to the ground.

There is always the contamination problem to be faced. I have already touched upon this: we have a unique opportunity to study Mars in its 'mint condition', and once we introduce any Earth life-forms which might be able to survive there and spread there is bound to be uncertainty. We could never be sure which organisms were genuinely Martian, and which had been taken there by our probes. Sterilization regulations are extremely strict, and are almost certain to be one hundred per cent. effective, but we must remember that Mars is not like the Moon; it has an atmosphere, so that it would be much more prone to widespread contamination.

The chances of bringing back anything harmful from Mars are remarkably slim, but they are not nil, and public opinion alone would ensure that the greatest precautions would be taken. This is why I imagine that when the first samples come back, they will be first examined either in a space-station or else in a Lunar Base—probably the former, because I anticipate that the samples will be available before any permanent base on the Moon has been set up.

Sample collection ought to clear up the life problem once and for all. With any luck, we may also find out whether there is ice under the surface of Mars. If so, then the problems for future astronauts will be considerably eased, because they will have all the water they require. This could be essential, because a request for essential supplies could not be met quickly; no space-craft will be able to do the Earth–Mars journey in less than several weeks, at least until we can tap some source of power as yet unknown to us.

I am not optimistic about sending astronauts to Mars in rockets of the kind used in the Apollo missions. Chemical fuels are not satisfactory, and we use them at present only because we have nothing better. With chemical propellants, we are forced to use the Sun's gravity and swing the space-ship out to Mars in a transfer orbit, so that it can 'coast' almost all the way without using any power at all. This involves a journey lasting for months. Add a period spent on Mars itself before it and the Earth are suitably placed for the return, plus the trip home, and the astronauts will have to be away for a couple of years. Since they will have to take all their supplies with them, the difficulties are painfully obvious. Provisions for a mission to the Moon are easily stored, but for a voyage to Mars the so-called minor problems, such as air supply, water, food and so on, become anything but minor.

One solution, favoured by some authorities but by no means all, is to use nuclear-powered rockets. These are not likely to go much faster, because their thrust will be low, but there should be no real limitation upon the amount of propellant to be carried, so that the distance can be cut down; there will be no need to follow a transfer orbit taking the probe half-way round the Sun. In other words, it may be possible to take short cuts, so reducing the time of travel. It is too early to say when this will become practicable, though it might be within the next fifty years.

We must also make up our minds whether a manned trip to Mars will be worthwhile. Objectors will argue that everything we want to know can be discovered from analysis of samples collected by automatic vehicles, and it is certainly true that the first astronauts to go to Mars will be running tremendous risks. Remember the Apollo 13 crisis, when the mission was cut short and the crew members were brought home ahead of time; there was no problem of food or water. Unfortunately, a Mars voyage cannot be truncated. If anything should damage a manned vehicle during its interplanetary flight, the outlook would be bleak, and accidents do happen. The Apollo 13 near-disaster was due to human error, but there are natural hazards too. To give only one example: at some time in the future we must, I am afraid, expect at least a few fatalities due to collisions between meteoroids and space-craft—and Mars is not so very far from the innermost part of the asteroid belt.

We know that astronauts (and cosmonauts) are not harmed by radiation so long as they remain aloft for no more than a few months. The record, set by the final crew of Skylab, exceeds eighty days. Whether a much longer exposure to cosmic rays and the like will be harmful remains to be seen. The general consensus of opinion seems to be that the risk is slight, but a nagging doubt remains.

The worst drawback about Mars is its lack of useful atmosphere. The carbon dioxide which makes up 95 per cent. of the whole is of little direct help, and in any case the Martian 'air' is so thin that it may not provide adequate protection against lethal radiations (though it is quite dense enough to cope with most incoming meteoroids). Suggestions have been made that Mars might be provided with an Earth-type atmosphere, so that astronauts could walk around unprotected. The practical problems of such a plan are terrifying, and we must also remember that Mars has a weak gravitational pull, so that an Earth-type atmospheric mantle would be slowly but inexorably lost. True, the wastage would be gradual, but it would happen.

If I were asked how such an atmosphere could be given to Mars, or how it could be prevented from escaping, I could only say that I do not know—though I hasten to add that I am neither an engineer nor a chemist!

With its present-day atmosphere, Mars does not provide the environment needed for the cultivation of what we may term useful plants. In other words, food cannot be produced there by natural methods, except inside airtight bases. Moreover, an astronaut who ventures outside his space-ship or his base will have to wear fully pressurized suiting, just as is necessary on the Moon, simply because the atmospheric density is so low that it corresponds to what we usually regard as a vacuum. Any rent in the wearer's suit would have prompt and extremely unpleasant consequences.

So far as we are concerned, life on Mars will always have to be maintained in a highly artificial environment. It is also true to say that even the first pioneers will have to set up a base of some sort or other, because they will have to spend some time there, and there is no hope of a rapid there-and-back Apollo-type reconnaissance. The popular, science-fiction form of base takes the shape of a hemispherical dome, kept inflated by the pressure of the atmosphere inside it, and equipped with elaborate airlocks to make sure that nothing leaks out when an astronaut is entering or leaving. Whether this idea will prove to be even remotely like the real thing remains to be seen, but there is nothing outrageous or illogical about it. (On the other hand, let us recall that the current space-stations, Skylab and Salyut, are quite different from the elegant wheel-shaped structures designed in the 1950s.)

As yet there is little point in making further speculations about the form of a base, because we still do not have enough information about Mars itself. In particular, we do not know whether there is an adequate supply of H_2O locked up in or below the surface. But the scientific uses of a base would be considerable, and the whole project might well lead on to true co-operation between the technically-advanced nations of the world. No Martian base can be American, Russian, British or Chinese. It must be genuinely international, and if we fail to reach this degree of mutual help we will certainly fail also to send men to Mars.

Incidentally, it seems most unlikely that the first trip will be made by a solo vehicle, as happened with the Moon. I envisage a whole 'Mars fleet', made up of at least three spaceships and probably more. But again we are looking so far ahead that judgement must, for the moment, be reserved.

There will be no communications problems between Earth and a Martian base; the Mariners and Vikings have shown that, though it is true that there will be times when direct contact is cut off (as happened with the first two Vikings between mid-November and late December 1976). On Mars itself, wheeled vehicles can presumably operate, and eventually —if several bases are set up—there is no reason why a sort of railway system should not be constructed (no doubt to work efficiently until some Martian Government is ill-advised enough to nationalize it). Air travel will be more restricted, because the tenuous atmosphere is of no use to conventional aeroplanes, and it seems that the rocket principle will have to be employed all the time.

The Martian ionosphere may be too ineffective to bounce back radio waves in the same way that ours does, but this does not mean that radio contact between two widely-separated bases will be impossible, because Phobos and Deimos can help. A relay transmitter on a satellite will be invaluable, and should present no problems whatsoever to a technology able to send men to Mars in the first place. Actually, I suggest that Deimos will be the more useful of the two, because it remains above the horizon, as seen from any one place on Mars, for more than two sols consecutively. Only from very high latitudes will the satellite link be lost, and I doubt whether the first bases will be set up in the Martian polar regions. (In any case, it will also be possible to launch artificial satellites to act as relays if Phobos and Deimos do not come up to expectations.)

Obviously the bases will have to be largely self-supporting. Whether they can ever become completely so is another question which cannot yet be answered, but there is a relevant point which I have made and which has caused a certain amount of discussion. How will human life on Mars evolve?

It would be absurd to suggest that the Martian expeditions

will be all-male, because the pioneers will have to spend protracted periods on the planet, and a spell of duty there will last for a long time. If I am right in suggesting that there will be thriving colonies by, say, A.D. 2200 (perhaps well before), children will be born on Mars and will grow up there. The surface gravity is only one-third of its Earth value, and we have to ask ourselves whether a child growing up on Mars will ever be able to adapt to the higher gravity of Earth. I think that there is a serious doubt about this, in which case we may reach the situation of having men and women who can never come to the planet of their forbears. They will have to remain on Mars, in space, or upon some world where the surface gravity is at an acceptable level for them. We will end up with two distinct species of *homo sapiens*, and it is conceivable that physical differences will develop over several generations, so that a glance will show which man or woman is a 'terrestrial' and which is a 'Martian'. This may sound far-fetched enough to be classed as science fiction. I agree. And yet I believe that it may have become cold fact within a very few centuries from now.

There is another suggestion which is definitely out of court. On Earth, our most serious potential danger is that of over-population, and unless there is either a world war, a natural disaster or an unintentional man-made hazard (such as an outbreak of disease which cannot be controlled), our planet is going to become extremely crowded. Idealists have maintained that the solution is to send our surplus population to the Moon or Mars, but they forget that life on these worlds must always be under artificial conditions, which imposes an immediate limitation. I am quite ready to believe that by A.D. 2500 there may be a permanent population of a million people on the Moon and ten million on Mars, but this is, metaphorically, no more than a drop in the ocean. It would be rather like trying to solve London's traffic problem by removing all the cars registered in Bognor Regis. Something must certainly be done before it is too late, but the Moon and Mars can give no substantial help, The whole problem is social rather than scientific, so that I am neither willing nor competent to discuss it here.

I know, only too well, that in these last few pages I have ventured into the realm of speculation, and I will end by returning to what we call sober science. Our knowledge of Mars has increased beyond all expectations since that great moment in 1965 when Mariner 4 sent back the first close-range pictures. Old theories have been rejected, and a few have been restored; we have found that even though Mars can support little or nothing in the form of natural life, it is a world of intense interest. With its giant volcanoes, its deep valleys, its craters, its dried-up rivers and its miniature moons, it is unique. Within the next few decades we will learn even more, and then we will be able to make up our minds just when the first astronauts will go there, to stand in the rocky deserts and look up at the redness of the sky. Mariner and Viking have blazed the trail, and the outlook is exciting. In the time of our grandchildren or great-grandchildren, the 'new Martians' will arrive.

Appendix I

OBSERVING MARS

MARS IS A DIFFICULT telescopic object. This may not seem easy to believe at first sight, because Mars, as we know, can become brighter than any other planet apart from Venus; but it is a fact. Anyone who buys a small telescope, pokes it through a bedroom window and expects to see a network of canals is doomed to disappointment.

Even when it is at its closest to us, the apparent diameter is a mere 25·7 seconds of arc, which is only about half that of Jupiter. And when Mars is at its greatest distance, the apparent diameter shrinks to 3·5 seconds of arc, comparable with that of remote Uranus. Generally speaking, little can be seen on the disk when the diameter drops below 6 or 7 seconds of arc—and at all times a telescope of considerable aperture is needed for any scientific work which may be regarded as useful. Mars is one planet upon which a high magnification has to be used, though this does not mean that one should go too far; a smaller, sharp image is always better than a large but blurred view.

When Mars is reasonably near opposition, a telescope as small as a 3-inch refractor will show the main markings: the polar caps (or, to be more precise, whichever polar cap happens to be turned toward us) and dark areas such as Syrtis Major and Acadalia Planitia. Even here there must be a qualification, because the polar cap shrinks rapidly with the onset of Martian spring and summer, and there are long periods when it cannot be seen at all except with large instruments. It may sometimes seem to vanish altogether. And, of course, the great dust-storms, such as that of 1971, conceal all surface details no matter what telescope is used.

For useful work, I consider that the minimum telescopic aperture is 8 inches for a reflector and 5 inches for a refractor. Other observers will disagree with me, and will say that a 6-inch reflector is adequate. They may well be right, but all I can do is to give a personal opinion. So far as magnification is concerned, I suggest that the proper procedure is to select the highest power which will give a really sharp image.

'Useful work' is a term which needs to be defined with regard to Mars. Now that we have the close-range Mariner 9 and Viking pictures, there is little point in setting out to do ordinary mapping; the situation is much the same as for the Moon, where the pre-Space Age observers concentrated upon cartography, whereas the modern amateurs (and some professionals) are searching for what we may call time-dependent phenomena. There is, however, one very important difference. The Moon's structural features do not change, and have not done so for many millions of years. Mars has an atmosphere, and this atmosphere is dusty; the dark areas do show alterations in outline and intensity which are not yet fully understood. Frankly, they are less easy to explain than they used to be in the days when it was still thought that the dark regions were due to vegetation. Windblown material seems to be the only logical answer to the problem, but further observations are needed, and this is where the amateur can join in—provided that he has adequate telescopic equipment.

The Syrtis Major is a case in point. Sometimes it is said to look narrow, sometimes broader; and though many of these alleged changes can be put down to errors in observation or interpretation, it is at least possible that they have a basis of reality. Then there is Hellas, which can sometimes appear so brilliant that it is easily mistaken for an extra ice-cap, but at other oppositions is so obscure that it is hard to identify at all. Argyre I shows variations in brightness of the same kind, though they are less striking.

Clouds are seen frequently, and observations are genuinely useful. Small clouds, occasionally well enough defined to have their positions determined, can shift from night to night, and can provide information about Martian wind velocities. The major dust-storms come into a different category, and here the best procedure is to do one's best to define their limits. They can spread with amazing rapidity, as happened in 1973. I observed the planet on 12 October, and the surface details were perfectly clear. The next three nights were cloudy. When I next observed, on 16 October, the dust had covered the planet, and I could see practically nothing at all.

Much depends upon the altitude of Mars. When the planet is low down a high magnification will be useless, and there is no alternative but to wait until the alternative has climbed to a respectable value. Also, it often happens that a very clear, transparent night will be unfavourable, with the disk of Mars wobbling like a jelly. Oddly enough, slight mist is sometimes advantageous, even though it cuts down the total light received, because the image can be pleasingly steady.

Without wanting to sound depressing, I must again stress that small telescopes are unable to show much on Mars even near opposition. There have been many published drawings made with such instruments, often showing fine details together with canals—all of which are completely spurious. This is no slur upon the integrity of the observers concerned, but the human eye is notoriously easy to deceive, and there is always the tendency to draw what one half-expects to see. Neither is this tendency confined to amateurs: far from it. One has only to look at the maps produced by Schiaparelli, Lowell and others, who were using powerful telescopes, but who still recorded the network of canals which we now know to be absolutely non-existent.

The phase should never be neglected when making a sketch. The fraction of the illuminated disk presented to us can go down to as little as 85%, and if the observer merely draws a circle and fills in the detail he can see there are bound to be major errors. Drawing the disk to the correct phase can, however, be a tedious process. Personally, I admit that I 'cheat' by using prepared disks of the type shown here (Fig. 45). They have the advantage of looking much neater than freehand phase drawings, and they are also shown against a black background, though I admit that this is not strictly necessary. To achieve absolute accuracy one ought to have a full set of disks, from 100% (i.e. a perfect circle, as at opposition) down to 85% (the minimum phase), but in practice I doubt whether one needs more than half a dozen—say 100%, 98, 95, 92, 89, and 86. An error of one or two per cent. is, frankly, unimportant.

In general it is wise to adopt a set scale for drawings; the Mars Section of the British Astronomical Association workers use 2 inches to the planet's full diameter, and this seems very suitable. Some observers vary the scale, drawing Mars largest

90%

93%

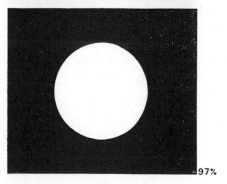

97%

Fig. 45. Disks to help in making drawings of Mars: phase 90 per cent., 93 per cent. and 97 per cent

when at opposition, but this seems an unnecessary refinement.

When sketching Mars, the first step is to survey the planet and see just what is on view. Then draw in the obvious details such as the polar caps and the main dark areas. When this has been done, check carefully and note the time (using the 24-hour clock, and ignoring Summer Time; always give G.M.T.). These main features should then be left unaltered. There is good reason for not changing them: Mars is rotating all the time, and the drift of the markings across the disk is perceptible even over periods of a few minutes.

Now change to the highest possible magnification and fill in the minor details, paying particular attention to the relative intensities of the various features, and concentrating upon anything which seems unusual (clouds in particular). Written notes, dealing with features of special interest, can be added, after which the whole drawing should be re-checked for accuracy. Finally, add the following data: date, time, name of observer, type and aperture of telescope, magnification, seeing conditions, and the longitude of the central meridian of Mars. Should any of these facts be omitted, the drawing promptly loses most of its value.

Seeing is usually given on the scale proposed by Antoniadi, ranging from I (perfect) through II (good), III (fair), IV (rather poor) down to V (so bad that one would not make a drawing at all unless there was some special reason for attempting it).

The longitude of the central meridian can be calculated easily, and involves nothing more frightening than simple addition or subtraction. Various publications, such as the *Handbook* of the British Astronomical Association, give the longitude of the central meridian for o hours G.M.T. each day, so that all one has to do is to allow for the interval between o hours and the time of observation. The longitude changes by the following amounts:

Hours	Change, degrees	Minutes	Change, degrees
1	14·6	1	0·2
2	29·2	2	0·5
3	43·9	3	0·7
4	58·5	4	1·0
5	73·1	5	1·2
6	87·7	6	1·5
7	102·3	7	1·7
8	117·0	8	2·0
9	131·6	9	2·2
10	146·2	10	2·4
		20	4·9
		30	7·3
		40	9·7
		50	12·2

Let me give a couple of examples from my own notebook.

(1). I observed Mars on 2 September 1975, at 01·20 G.M.T., 15 in. reflector × 360; seeing III, phase 85%. Looking up the *Handbook*, I found that at o hours on 2 September the central longitude was 192·5.

Longitude at o hours:	092·5
+ 1 hour	14·6
+ 20 minutes	4·9
	—
Longitude at 01·20	112·0

This meant that the Syrtis Major was on the far side of the disk; but features such as the Mare Sirenum were on view— though since the diameter of the disk was only 9 seconds of arc, and the seeing was not good, I could make out little detail.

(2). Observation: 3 January 1976, at 16·27 G.M.T.: 15-in. refl. × 400, seeing II.

This time we have to subtract, because the observation was made 7 hours 33 minutes before o hours G.M.T.

Longitude of central meridian at o hours on 4 January: 036·7.

We have to subtract for 7h 33m.	7 hours:	102·3
	+30 minutes	7·3
	+ 3 minutes	0·7
		110·3

Since one cannot take 110·3 from 036·7, we must start by adding 360 degrees. 360 + 036·7 = 396·7.

Longitude at o hours on 4 January:	396·7
Subtract for 7 h 33 m.:	110·3
Longitude at 16·27 on 3 January:	286·4

Therefore the Syrtis Major, whose longitude is approximately 290 degrees, was almost on the central meridian.

Because unconscious prejudice is the observer's greatest enemy, particularly when dealing with Mars, it is usually a good idea to leave working out the longitude of the central meridian until after the observation has been made—though I agree that this is rather pointless when some unmistakable feature is on view. In any case, avoid drawing any details which cannot be seen with absolute certainty.

I will not pretend that the average amateur can make many really valuable contributions to the study of Mars, but opportunities do occur sometimes—and in any event there is a tremendous amount of personal pleasure to be drawn from looking at the planet and recording what one can see.

Appendix II

OPPOSITIONS OF MARS

The interval between successive oppositions of Mars is not constant; it may be as much as 810 days or as little as 764 days. Oppositions between 1900 and 1956 occurred on the following dates:

1901	Feb. 22	1928	Dec. 21
1903	Mar. 29	1931	Jan. 27
1905	May 8	1933	Mar. 1
1907	July 6	1935	Apr. 6
1909	Sept. 24	1937	May 19
1911	Nov. 25	1939	July 23
1914	Jan. 5	1941	Oct. 10
1916	Feb. 9	1943	Dec. 5
1918	Mar. 15	1946	Jan. 13
1920	Apr. 21	1948	Feb. 17
1922	June 10	1950	Mar. 23
1924	Aug. 23	1952	Apr. 30
1926	Nov. 4	1954	June 24

Opposition dates, between 1954 and 1981:

Date		Minimum distance from Earth. Millions of		Maximum magnitude	Maximum apparent diameter, seconds of arc
		Miles	Kilometres		
1956	Sept. 11	35·2	56·6	−2·6	24·7
1958	Nov. 17	45·4	73·0	−1·9	18·9
1960	Dec. 30	56·3	90·6	−1·3	15·4
1963	Feb. 4	62·2	100·1	−1·0	14·0
1965	Mar. 9	62·0	99·8	−0·9	14·0
1967	Apr. 15	55·8	89·8	−1·3	15·6
1969	May 31	44·5	71·7	−2·0	19·3
1971	Aug. 10	34·9	56·2	−2·7	24·9
1973	Oct. 25	40·4	65·0	−2·2	21·1
1975	Dec. 15	52·4	84·3	−1·5	16·2
1978	Jan. 22	60·8	97·8	−1·1	14·3
1980	Feb. 25	63·2	101·7	−0·8	13·8

Opposition dates between 1981 and 2000 will be:

Date	Maximum diameter, seconds of arc
1982 Mar. 31	14·7
1984 May 11	17·4
1986 July 10	22·1
1988 Sept. 28	23·6
1990 Nov. 27	17·8
1993 Jan. 7	14·0
1995 Feb. 12	13·8
1997 Mar. 17	14·2
1999 Apr. 24	16·2

The date of closest approach may differ from the opposition date by as much as ten days. The following list gives the closest approaches of Mars between the memorable year 1877 and the end of our own century:

Date	Minimum distance from Earth, millions of	
	Miles	Kilometres
1877 Sept. 5	34·8	56·0
1892 Aug. 26	34·5	55·5
1909 Sept. 18	36·2	58·3
1924 Aug. 22	34·5	55·5
1939 July 23	36·1	58·0
1956 Sept. 11	35·2	56·6
1971 Aug. 10	34·9	56·2
1988 Sept. 28	36·3	58·4

At its greatest distance from Earth, at superior conjunction, Mars may recede to as much as 249 million miles (400 million km). The dates of superior conjunction for the period 1960–2000 are as follows:

1961	Dec. 14	1972	Sept. 7	1983	June 4	1993	Dec. 27
1964	Feb. 17	1974	Oct. 16	1985	July 18	1996	Mar. 6
1966	Apr. 29	1976	Nov. 26	1987	Aug. 27	1998	May 13
1968	June 21	1979	Jan. 21	1989	Oct. 1	2000	July 2
1970	Aug. 2	1981	Apr. 2	1991	Nov. 10		

Appendix III

NUMERICAL DATA

Distance from the Sun:
maximum	154,861,000 miles (1·666 a.u.)
mean	141,637,000 miles (1·524 a.u.)
minimum	128,412,000 miles (1·381 a.u.)

Axial rotation period (sol): 24 hours 37 minutes 22·7 seconds
Sidereal period: 686·980 days, or 668·60 sols
Orbital velocity, miles per
second: maximum   16·5
              mean   15·0
              minimum 13·6
Orbital inclination: $1°50'59''·4$ ($= 1°·9$)
Orbital eccentricity: 0·093
Mean synodic period: 779·74 days
Mean sidereal motion in
24 hours: 1886·52 seconds of arc
Axial inclination: $23°59'$
Diameter: 4219 miles
Apparent diameter: maximum $25''·7$, minimum $3''·5$
Maximum magnitude: −2·8
Albedo (mean): 0·16
Volume (Earth = 1): 0·1504
Mass (Earth = 1): 0·1074
Density (water = 1): 3·94
Surface area (Earth = 1): 0·28
Surface gravity (Earth = 1): 0·3799
Polar compression: 0·01
Escape velocity: 3·1 miles per second

Satellite Data

	Phobos	Deimos
Discoverer:	Hall, 16 Aug. 1877	Hall, 11 Aug. 1877
Mean distance from centre of Mars:	5825 miles	14,575 miles
Sidereal period:	7h 39m 13s·85 (0·319 days)	30h 17m 54s·87 (1·262 days)
Mean synodic period:	7h 39m 26s·6	30h 21m 15s·7
Orbital inclination:	1·1 degrees	1·8 degrees

195

Orbital eccentricity:	0·0210	0·0028
Mean angular distance from Mars, at mean opposition:	24·6 seconds of arc	61·8 seconds of arc
Mean visual opposition magnitude:	11·6	12·8
Diameter in miles (mean):	14	8

MARS PROBES, 1962–1977

Probe	Launch date	Arrival date	Closest approach	Weight (lb.)	Objectives	Remarks
Mars 1	1962 Nov. 1	?	?	1965	Fly-by	Contact broken at 65,900,000 miles.
Mariner 3	1964 Nov. 5	—	—	575	Fly-by	Shroud failure. Entered solar orbit, but went nowhere near Mars.
Mariner 4	1964 Nov. 28	1965 July 14	6118 miles	575	Fly-by	Returned 21 pictures of Mars. Now in solar orbit. Contact finally lost on 20 December 1967.
Zond 2	1964 Nov. 30	?	?	?	Fly-by	Contact lost. Probable rendezvous with Mars in August 1965.
Mariner 6	1969 Feb. 24	1969 July 31	2120 miles	910	Fly-by	Flew over Martian equator, returning 75 pictures. Now in solar orbit.
Mariner 7	1969 Mar. 27	1969 Aug. 4	2190 miles	910	Fly-by	Flew over southern part of Mars, returning 126 pictures. Now in solar orbit.
Mariner 8	1971 May 8	—	—	2150	Mars orbiter	Total failure: fell in the sea!
Mars 2	1971 May 19	1971 Nov. 27	Lander+orbiter	10,250	Studies from orbit and surface	In Mars orbit, 1530 × 15,250 miles. Dropped a capsule on to Mars, carrying a Soviet pennant: lat. 44°·2, S long. 213°·2 E.
Mars 3	1971 May 28	1971 Dec. 2	Lander+orbiter	10,250	Studies from orbit and surface	In Mars orbit, 970 × 133,000 miles. Lander touched down at lat. 45°, S long. 158°, but lost contact after 20 seconds: no useful data received.

Name	Launch date	Orbit/arrival date	Vehicle	Weight	Mission	Remarks
Mariner 9	1971 May 30	1971 Nov. 13 (in Mars orbit)	1025 miles	2150	Orbiter	In Mars orbit, 1025 × 10,500 miles. Operated from 13 Nov. 1971 to 27 Oct. 1972, returning 7329 pictures.
Mars 4	1973 July 21	1974 Feb. 10	?	?	Orbiter	Failed to orbit. Missed Mars by over 1300 miles.
Mars 5	1973 July 25	1974 Feb. 12	?	?	Orbiter	In Mars orbit. Contact lost.
Mars 6	1973 Aug. 5	1974 Mar. 12	Lander+orbiter	?	Studies from orbit and surface	Contact lost during landing sequence. Probable landing on Mars at lat. 24° S, long. 25° W.
Mars 7	1973 Aug. 9	1974 Mar. 9	Lander+orbiter	?	Studies from orbit and surface	Failed to orbit; missed Mars by 800 miles.
Viking 1	1975 Aug. 20	1976 June 19 (in Mars orbit)	Lander+orbiter	7700	Studies from orbit and surface	Landed, 20 July 1976, in Chryse (lat. 22°.4 N, long. 47°.5 W).
Viking 2	1975 Sept. 9	1976 Aug. 7 (in Mars orbit)	Lander+orbiter	7700	Studies from orbit and surface	Landed, 3 Sept. 1976, in Utopia (lat. 48° N, long. 226° W).

The Mariners and Vikings are American; the Mars and Zond vehicles, Russian.

Appendix V

NAMED FEATURES ON MARS

These names have been adopted by the International Astronomical Union. The list is complete to 1977, but will no doubt be extended in the very near future. Extra names which are already in common use (e.g. Yuty, the prominent 11-mile crater in Chryse) will certainly receive official approval before long.

	Longitude degrees	Latitude degrees
Catena (chain of craters)		
Coprates	66 to 56	−15
Ganges	71 to 67	−02 to −03
Tithonia	98 to 80	−06
Chasma (canyon)		
Australis	270	−80 to −88
Borealis	65 to 30	+85
Candor	78 to 73	−04 to −06
Capri	52 to 32	−14 to −03
Coprates	68 to 54	−11 to −14
Eos	51 to 32	−16 to −17
Ganges	52 to 48	−08
Hebes	81 to 73	+01 to −01
Ius	98 to 80	−07
Juventæ	61	−04
Melas	78 to 70	−08 to −12
Ophir	77 to 64	−03 to −09
Tithonia	90 to 80	−04
Dorsum (ridge)		
Argyre	70	−61 to −65
Fossa (long, narrow valley)		
Alba	117 to 109	+38 to +49
Cerauniæ	107	+25
Claritas	108 to 105	−19 to −32
Elysium	225 to 219	+28 to +26
Hephæstus	240 to 233	+22 to +18

	Longitude degrees	Latitude degrees
Mareotis	85 to 69	+41 to +48
Medusæ	162	−08
Memnonia	158 to 140	−22 to −15
Nili	284 to 279	+20 to +26
Sirenum	163 to 138	−36 to −27
Tantalus	105 to 99	+34 to +47
Tempe	80 to 62	+35 to +46
Thaumasia	100 to 80	−36 to −40

Labyrinthus (valley complex)

Noctis	110 to 92	−05 to −08

Mansa (flat-topped elevation)

Deuteronilus	346 to 340	+42 to +45
Nilosyrtis	290	+32
Protonilus	315	+38

Mons (mountain)

Arsia	121	−09
Ascræus	104	+12
Elysium	213	+25
Olympus	133	+18
Pavonis	113	+01

Montes (mountains)

Charitum	50 to 32	−57
Hellesponti	315	−45 to −48
Nereidum	57 to 43	−48 to −38
Phlegra	195	+31 to +46
Tharsis	125 to 101	−12 to +16

Planitia (plain)

Acidalia	30	+48
Arcadia	155	+48
Amazonis	160	+13
Argyre	43	−49
Chryse	45	+17
Elysium	210	+15
Hellas	290	−45
Isidis	270	+15
Syrtis	290	+15
Utopia	235	+35

Planum (plateau or high plain)	Longitude *degrees*	Latitude *degrees*
Auroræ	52 to 48	−10 to −11
Hesperia	258 to 242	−10 to −35
Lunæ	70 to 60	+05 to +20
Ophir	61 to 55	−09 to −12
Solis	98 to 88	−20 to −30
Syria	105 to 100	−10 to −18
Sinai	90 to 70	−10 to −20

Patera (shallow, scalloped crater)		
Alba	110	+40
Amphitrites	299	−59
Apollinaris	186	−08
Biblis	124	+02
Hadriaca	267	−31
Orcus	181	+14
Pavonis	121	+03
Tyrrhena	253	−22
Uranius	93	+26

Tholus (hill)		
Albor	210	+19
Australis	323	−57
Ceraunius	97	+24
Hecates	210	+32
Hippalus	89	+76
Iaxartes	15	+72
Jovis	117	+18
Kison	358	+73
Ortygia	8	+70
Tharsis	91	+14
Uranius	98	+26

Vallis (valley)		
Al Qahira	202 to 194	−23 to −15
Ares	23 to 14	+02 to +10
Auqakuh	298	+28
Huo Hsing	295 to 292	+32 to +28
Maadim	183	−27 to −20
Mangala	151	−10 to −4
Marineris	95 to 45	−05 to −15
Nirgal	44 to 36	−32 to −27

	Longitude *degrees*	Latitude *degrees*
Kasei	70 to 56	+21
Shalbatana	45	+01 to +15
Simud	40 to 37	00 to +14
Tiu	32	+10 to +18

Vastitas (widespread lowland)
Borealis	continuous	+55 to +67

	Long.	Lat.		Long.	Lat.
Adams, W. S.	197	+31	Darwin, G. H. & C.	20	−57
Agassiz, J. L. R.	89	−70	Dawes, W. R.	322	−9
Airy, G. B.	0	−0.5	Denning, W. F.	326	−18
Antoniadi, E. M.	299	+22	Douglass, A. E.	70	−52
Arago, F.	330	+10	Du Martheray, M.	266	−6
Arrhenius, S.	237	−40	Du Toit, A. L.	46	−72
Bakhuysen,			Eddie, L. A.	218	+12
H. G. van de S.	344	−23	Eiriksson, L.	174	−19
Baldet, F.	295	+23	Escalante, F.	245	0
Barabashov, N.	69	+47	Eudoxus	147	−44
Barnard, E. E.	298	−61	Fesenkov, V. G.	87	+22
Becquerel, H.	8	+22	Flammarion, C.	312	+26
Beer, W.	8	−15	Flaugergues, H.	341	−17
Bianchini, F.	97	−64	Focas, J. H.	347	+34
Bjerknes, W.	189	−43	Fontana, F.	73	−64
Boeddicker, O.	197	−15	Fournier, G. & V.	287	−4
Bond, G. P.	36	−33	Gale, W. F.	222	−6
Bouguer, P.	333	−19	Galilei, G.	27	+6
Brashear, J. A.	120	−54	Galle, J. G.	31	−51
Briault, P.	270	−10	Gilbert, G.	274	−68
Burroughs, E. R.	243	−72	Gill, D.	354	+16
Burton, C. E.	156	−14	Gledhill, J.	273	−53
Campbell,			Graff, K.	206	−21
W. W. & J. W.	195	−54	Green, N. E.	8	−52
Cassini, J. D.	328	+24	Hadley, G.	203	−19
Cerulli, V.	338	+32	Haldane, J. B.	231	−53
Chamberlain, T. C.	124	−66	Hale, G. E.	36	−36
Charlier, C. V. L.	169	−69	Halley, E.	59	−49
Clark, A.	134	−56	Hartwig, E.	16	−39
Coblentz, W. W.	91	−55	Heaviside, O.	95	−71
Columbus, C.	166	−29	Helmholtz, H. von	21	−46
Comas Solá, J.	158	−20	Henry, P. & P.	336	+11
Copernicus, N.	169	−50	Herschel, W. & J.	230	−14
Crommelin,			Hipparchus	151	−44
A. C. D.	10	+5	Holden, E. S.	34	−26
Cruls, L.	197	−43	Holmes, A.	292	−75
Curie, P.	5	+29	Hooke, R.	44	−45
Daly, R. A.	22	−66	Huggins, W.	204	−49
Dana, J. D.	32	−73	Hussey, T. J.	127	−54

	Long.	Lat.		Long.	Lat.
Hutton, J.	255	−72	Milankovitch, M.	147	+55
Huxley, T. H.	259	−63	Millochau, G.	275	−21
Huygens, C.	304	−14	Mitchel, O. M.	284	−68
Janssen, P. J. C.	322	+3	Molesworth, P. B.	211	−28
Jerry-Desloges, R.	276	−9	Moreux, T.	315	+42
Jeans, J.	206	−70	Müller, G. & H. J.	232	−26
Joly, J.	42	−75	Nansen, F.	141	−50
Jones, H. Spencer	20	−19	Newcomb, S.	358	−24
Kaiser, F.	340	−46	Newton, I.	158	−40
von Kármán, T.	59	−64	Nicholson, S. B.	166	0
Keeler, J. E.	152	−61	Niesten, L.	302	−28
Kepler, J.	219	−47	Oudemans,		
Knobel, E.	226	−6	J. A. C.	92	−10
Korolev, S. P.	196	+73	Pasteur, L.	335	+19
Kuiper, G. P.	157	−57	Perepelkin, E. J.	65	+52
Kunowsky, G. K. F.	9	+57	Peridier, J.	276	+26
Lambert, J. H.	335	−20	Pettit, E. E.	174	+12
Lamont, J.	114	−59	Phillips, T. & J.	45	−67
Lampland, C. O.	79	−36	Pickering, W. & E.	133	−34
Lassell, W.	63	−21	Playfair, R.	125	−78
Lau, H. E.	107	−74	Porter, R. W.	114	−50
Le Verrier, U. J. J.	343	−38	Priestley, J.	228	−54
Liais, E.	253	−75	Proctor, R. A.	330	−48
Li Fan	153	−47	Ptolemæus	158	−46
Liu Hsin	172	−53	Quénisset, F.	319	+34
Lockyer, N.	199	+28	Rabe, W.	325	−44
Lomonosov, M. V.	8	+65	Radau, R.	5	+17
Lowell, P.	81	−52	Rayleigh, J. W.	240	−76
Lyell, C.	15	−70	Redi, F.	267	−61
Lyot, B.	331	+50	Renaudot, G.	297	+42
Mädler, J. H. von	357	−11	Reuyl, D.	193	−10
Magelhæns	174	−32	Reynolds, O.	160	−74
Maggini, M.	350	+28	Richardson, L. F.	181	−73
Main, R.	310	−77	Ritchey, G. W. F.	51	−29
Maraldi, G.	32	−62	Ross, F. E.	108	−58
Mariner	164	−35	Rossby, G. G.	192	−48
Marth, A.	3	+13	Rudaux, L.	309	+38
Martz, E. P.	217	−34	Russell, H. N.	348	−55
Maunder, E. W.	358	−50	Rutherford, E.	11	+19
McLaughlin, D. B.	22	+22	Schaeberle, J. M.	310	−24
Mendel, G.	199	−59	Schiaparelli, G. V.	343	−3
Mie, G.	220	+48	Schmidt, J. & O.	79	−72

	Long.	Lat.		Long.	Lat.
Schröter, J. H.	304	−2	Trouvelot, E. L.	13	+16
Secchi, A.	258	−58	Trümpler, R. J.	151	−62
Sharonov, V. V.	59	+27	Tycho Brahe	214	−50
Sklodowska, M.	3	+34	Tyndall, J.	190	+40
Slipher,			Very, F. W.	177	−50
E. C. & V. M.	84	−48	da Vinci, L.	39	+2
Smith, W.	103	−66	Vinogradsky, S. N.	217	−56
South, J.	339	−77	Vishniac, W.	276	−77
Spallanzani, L.	273	−58	Vogel, H.	13	−37
Steno, N.	115	−68	Wallace, A. R.	249	−53
Stokes, G. G.	189	+56	Wegener, A.	4	−65
Stoney, G. J.	134	−69	Weinbaum, S.	245	−66
Suess, E.	179	−67	Wells, H. G.	238	−60
Teisserenc de Bort,			Williams, A. S.	164	−18
J.	315	+1	Wirtz, K.	26	−49
Terby, F.	286	−28	Wislencius, W.	349	−18
Tikhov, G. A.	254	−51	Wright, W. H.	151	−59

Craters on Phobos

D'Arrest, H.
Hall, A.
Roche, E.
Sharpless, B. P.

Stickney, A.
Todd, D.
Wendell, W.

(There is also one named ridge on Phobos: Kepler Dorsum.)

Craters on Deimos

Swift, J.

Voltaire, F. M. A.

INDEXES

INDEX TO FORMATIONS REFERRED TO IN
THE TEXT

209

GENERAL INDEX